ENVIRONMENTAL SCIENCE, ENGINEERING AND TECHNOLOGY

ENDOSULFAN

USES, TOXICOLOGICAL PROFILE AND REGULATION

ENVIRONMENTAL SCIENCE, ENGINEERING AND TECHNOLOGY

Additional books and e-books in this series can be found on Nova's website under the Series tab.

ENVIRONMENTAL SCIENCE, ENGINEERING AND TECHNOLOGY

ENDOSULFAN

USES, TOXICOLOGICAL PROFILE AND REGULATION

ISHWAR CHANDRA YADAV
AND
NINGOMBAM LINTHOINGAMBI DEVI
EDITORS

Copyright © 2019 by Nova Science Publishers, Inc.

All rights reserved. No part of this book may be reproduced, stored in a retrieval system or transmitted in any form or by any means: electronic, electrostatic, magnetic, tape, mechanical photocopying, recording or otherwise without the written permission of the Publisher.

We have partnered with Copyright Clearance Center to make it easy for you to obtain permissions to reuse content from this publication. Simply navigate to this publication's page on Nova's website and locate the "Get Permission" button below the title description. This button is linked directly to the title's permission page on copyright.com. Alternatively, you can visit copyright.com and search by title, ISBN, or ISSN.

For further questions about using the service on copyright.com, please contact:
Copyright Clearance Center
Phone: +1-(978) 750-8400   Fax: +1-(978) 750-4470   E-mail: info@copyright.com.

NOTICE TO THE READER

The Publisher has taken reasonable care in the preparation of this book, but makes no expressed or implied warranty of any kind and assumes no responsibility for any errors or omissions. No liability is assumed for incidental or consequential damages in connection with or arising out of information contained in this book. The Publisher shall not be liable for any special, consequential, or exemplary damages resulting, in whole or in part, from the readers' use of, or reliance upon, this material. Any parts of this book based on government reports are so indicated and copyright is claimed for those parts to the extent applicable to compilations of such works.

Independent verification should be sought for any data, advice or recommendations contained in this book. In addition, no responsibility is assumed by the Publisher for any injury and/or damage to persons or property arising from any methods, products, instructions, ideas or otherwise contained in this publication.

This publication is designed to provide accurate and authoritative information with regard to the subject matter covered herein. It is sold with the clear understanding that the Publisher is not engaged in rendering legal or any other professional services. If legal or any other expert assistance is required, the services of a competent person should be sought. FROM A DECLARATION OF PARTICIPANTS JOINTLY ADOPTED BY A COMMITTEE OF THE AMERICAN BAR ASSOCIATION AND A COMMITTEE OF PUBLISHERS.

Additional color graphics may be available in the e-book version of this book.

Library of Congress Cataloging-in-Publication Data

ISBN: 978-1-53615-910-3
Library of Congress Control Number:2019945917

Published by Nova Science Publishers, Inc. † New York

*To my whole family
who understand my passion
for insecticide and
supported all the time*

CONTENTS

Preface ix

Chapter 1 An Overview of Endosulfan: Characteristics, Eco-Toxicological Impact, and Regulatory Status   1
Ishwar Chandra Yadav and Ningombam Linthoingambi Devi

Chapter 2 Endosulfan Contamination in Indian Environment   25
Amrendra Kumar and Ningombam Linthoingambi Devi

Chapter 3 Endosulfan: Practices, Policies, and Eco-Toxicological Impacts in India   53
Sanjenbam Nirmala Khuman, Girija Bharat, Avanti Roy-Basu, Piyush Mohapatro and Paromita Chakraborty

Chapter 4 Endosulfan Evaluation in Human Samples   81
Virgínia Cruz Fernandes, Maria Luísa Correia-Sá, Maria Luz Maia, Cristina Delerue-Matos and Valentina Fernandes Domingues

Chapter 5	Residue Levels, Health Impacts, and Eco-Toxicity of Endosulfan in Tropical Environment *M. D. M. D. W. M. M. K. Yatawara*	**141**
Chapter 6	Status of Endosulfan Contamination and Management Practices in Pakistan *Mureed Kazim and Jabir Hussain Syed*	**175**
Chapter 7	Endosulfan Contamination in Soil: Sources, Impact and Bioremediation *Somenath Das, Anand Kumar Chaudhari, Ajay Kumar and Vipin Kumar Singh*	**205**
Chapter 8	Biodegradation and Detoxification of Chlorinated Pesticide Endosulfan by Soil Microbes *Arpan Mukherjee, Vipin Kumar Singh and Somenath Das*	**247**

About the Editor   273

Index   275

Related Nova Publications   291

Preface

Endosulfan is a semi-volatile organochlorine insecticide composed of two isomers (α- and β-) in the proportion of 70:30. It is persistent, bio-accumulative, and ubiquitous and is toxic to all species. It has caused widespread poisonings and suffering around the world. Hence, a global effort is needed to reduce or eliminate the menace of endosulfan.

In this respect, understanding the basic characteristics, eco-toxicological impact, distribution in the environment and its remedial measure are the important aspects for eliminating the risk of endosulfan and maintaining the healthy environmental for future. This book is intended for the researcher, students, toxicologist, health official and specialist in environmental science, ecotoxicology and health science. The content of the book is divided into 8 chapters, each of which discusses various aspects of endosulfan together with residual data of endosulfan from South Asian countries. The basic information about endosulfan is discussed in chapter 1. Chapter 2 and 3 describe the residual contamination, eco-toxicological impacts and regulatory policy of endosulfan in India environment. Chapter 4 relates to the evaluation of endosulfan in human samples. Chapters 5 and 6 focused on the residual level and management of endosulfan in tropical countries including Sri Lanka and Pakistan. The remediation techniques available for reducing the impact of endosulfan is detailed in chapter 7 and 8.

We are very optimistic that the reader of the book will definitely like the scientific content of the chapters to enhance their understanding of endosulfan. We would like to thank each contributor of the book for their considerable effort to present the most valuable accomplishment in their fields.

12 April, 2019

I.C Yadav
N.L. Devi

In: Endosulfan
Editors: I. C. Yadav and N. L. Devi
ISBN: 978-1-53615-910-3
© 2019 Nova Science Publishers, Inc.

Chapter 1

AN OVERVIEW OF ENDOSULFAN: CHARACTERISTICS, ECO-TOXICOLOGICAL IMPACT, AND REGULATORY STATUS

Ishwar Chandra Yadav[1,*]
and Ningombam Linthoingambi Devi[2]

[1]Department of International Environmental and Agricultural Science (IEAS), Tokyo University of Agriculture and Technology (TUAT) Tokyo, Japan
[2]Department of Environmental Sciences, Central University of South Bihar, Bihar, India

ABSTRACT

Endosulfan is a 'persistent organic pollutant' (POP) as defined under the Stockholm Convention. It is persistent in the environment, bio-accumulative, demonstrates long-range environmental transport, and causes adverse effects to human health and the environment. Endosulfan is listed as a POPs in the Convention on Long-range Transboundary Air

---

* Corresponding Author's Email: icyadav.bhu@gmail.com.

Pollution (LRTAP) and is recognized as a Persistent Toxic Substance (PTS) by the United Nations Environment Programme (UNEP). Although many of the organochlorine pesticides (OCPs) have been largely eliminated from their use in global agriculture following international recognition of their long term negative impacts on the global environment, endosulfan remains the major exception and is still widely applied to crops – particularly in the developing world. Due to its potential to evaporate and travel long distances in the atmosphere, endosulfan has become one of the world's most widespread pollutants. Endosulfan is very toxic to nearly all kinds of organisms. It is one of the most frequently reported causes of unintentional poisoning, particularly in Asia, Latin America, and West Africa. Most cases occur as a result of occupational exposure. Acute endosulfan poisoning can cause convulsions, psychiatric disturbances, epilepsy, paralysis, brain edema, impaired memory, and death. Long term exposure is linked to immunosuppression, neurological disorders, congenital birth defects, chromosomal abnormalities, mental retardation, impaired learning, and memory loss. This chapter deals with the physicochemical characteristics of endosulfan, their worldwide production and uses, health and environmental effects and fate in the environment. Further, the worldwide regulatory status of endosulfan has been also discussed.

Keywords: organochlorine, persistence, bioaccumulation, insecticides

1. INTRODUCTION

Endosulfan is a broad spectrum organochlorine insecticide and acaricide which act as a contact poison in a large group of insects and mites [1]. Endosulfan has been widely used in many parts of the world, including the European Union, India, Indonesia, Australia, Canada, United States, Mexico and Central America, Brazil and China [2-8]. It is effective against numerous types of insect pests and mites and has been used for 5 decades [9]. Thus, it is utilized on a large variety of crops such as grains, tea, fruits, and vegetables. Further, it is widely applied to non-food crops, for instance, tobacco and cotton, and as a wood preservative [1]. The emulsifiable concentrate (EC), wettable powder (WP), ultra-low volume (ULV) liquid, granules (G), dust (D) and smoke tablets are the different formulation of endosulfan available in the market. In the market, endosulfan is accessible with a number of trade names. Table 1 describes some of the trade names of the endosulfan used in the different country.

Table 1. Some trade name of endosulfane used in different country

SN	Country name	Trade name
1	Bangladesh	Thiodan
2	Brunei	Thiodan, Fezdion
3	Chile	Parmazol E, Flavylon, Galgofan, Thiodan, Thionex, Thionyly methofan
4	India	Agrosulphan, Agiro Sulphan, Banej Sulphan, Cilo Sulphan, Endo Sulphan, ESulfan, Endo Chithin, Endocid, Endonit, Endomil, Endosol, Endostar, Endosun, Endotaf, Endostan, Endocing, Endocide, Endosulpher, Gaydan, Gilnore Endorifan, Hexa-sulfan, Hildan, Hockey Endosulfan, Hy-sulfan, Kemu Sulfan, Hilexute-Sulfan, Krushi Endosulfan, Lusu Sulfan, Marvel-Micosulfan, Mico Thansulfan, Pary Sulfan, Pesticel, Remisfan, Sico sulfan, Solesulfan, Sujadin, Sulfan, Tej Sulfan, Thiodon, Thiokill, Thionel, Thionex, Thioton, Veg-fru Thiotox, Veg-fru Thiotex, Vika sulfan.
5	Indonesia	Thiodan, Fanodan, Dekasulfan
6	South Korea	Malix, Thiolix.
7	Pakistan	Siagon, Thiodan, Thioluxan
8	Philippines	Atlas Endosulfan, Endosulfan, Contra, Endox, Thiodan
9	Sri Lanka	Thiodan, Thionex, Endomack, Endocel, Baurs Endosulfan, Harcros Harcosan, Red Star Anglo-sulfan
10	Thailand	hiodan, A. B. Fan, Aggrodan, Agridan, Bensodan, Bensocarb, Beosit, Brook, Clement, Dew Dan, Dior 35, Dori, Dumpersan, E C Sulfan, Egodan, Endan, Endodan, Endosulfan, Endrew, Endye, Endyne, Etonic, Exxo-Z, Famcodan, Fortune, Freedan, Gardner, Gycin, Hor Mush, Hydrodan, J-teedan, Jack Dum, Kasidan, L P dan, Lordjim, Malix, Manyoo, Metrodan, Nayam, Newcodan, Nockdyne, Ox Xa, Patodan, Pestdye, Pro-d-dan, Sandan, Shevanex, Simadan, Sonydan, Summer, Tanadan, Teophos, Thanyacarb, Thimul, Thiofor, Torpidan, Urofen, Wephos, Zumic.
11	Other name	Chlorthiepin, Cyclodan, Endox, Thifor, Thiomul, Thionate.

Generally, air-blast or ground boom sprayers are used to spray the pesticide in the crop in temperate regions [10]. Commercially, endosulfan is made up of two stereoisomers, i.e., α- and β- endosulfan in the proportion of 70:30 [11-12]. It is lethal for the aquatic organism [13]. In mammals, it causes reproductive and neurological disorder [14, 15]. In soil, endosulfan is changed over into endosulfan sulfate by oxidation and doil endosulfan by hydrolysis. These two isomers have a dissimilar level of persistence in the environment. The half-life of total endosulfan is 1336 days, while α- and β-endosulfan may disappear/degrade after 27.5 and 157 days, respectively under aerobic condition. Endosulfan sulfate is long lasting and toxic in the environment than α- and β-endosulfan isomers.

On account of its semi-volatility and relative persistence, endosulfan is an omnipresent environmental contaminant that has been reported in the diverse group of environmental matrices. It has been accounted in the air, soil, water, and vegetation, frequently emission from the area of direct application. The Intergovernmental Forum on Chemical Safety (IFCS) distinguished endosulfan as an intensely lethal pesticide that can cause significant health problems for developing countries and economies on the move [16]. In 2002, the United States Environmental Protection Agency recommended that the exposure to endosulfan may result in both intense and incessant danger in terrestrial and aquatic environments.

2. Physical and Chemical Characteristics of Endosulfan

Endosulfan is a chlorinated hydrocarbon insecticide. In its pure form, endosulfan is a colorless crystalline solid while the technical grade comprised of crystalline flakes with cream to dark color and a faint odor of sulfur dioxide. Technical endosulfan is composed of at least 94% of two pure isomers, α- and β-endosulfan, while endosulfan sulfate is an oxidation product of technical endosulfan [17]. It is also found in the environment due to oxidation by biotransformation of endosulfan [18]. The two isomers of endosulfan are semi-volatile, with comparative vapor pressures to other chlorinated pesticides, thereby making them vulnerable to volatilization followed by atmospheric transport and deposition. The chemical name of endosulfan is 6,7,8,9,10,10-hexachloro-1,5,5a,6,9,9a-hexahydro- 6, 9-methano-2,3,4-benzadioxathiepin 3-oxide, with a molecular formula of $C_9H_6Cl_6O_3S$.

Table 2. Physical and chemical properties of endosulfan

Property	α-endosulfan	β-endosulfan	Endosulfan sulfate	References
Molecular weight	406.9 g/mol	406.9 g/mol	422.92 g/mol	Sarafin, 1987; Goerlitz, 1987; Shen and Wania 2005
CAS no.	959-98-8	3321-65-9	1031-07-8	
Molecular formula	$C_9H_6Cl_5O_3S$	$C_9H_6Cl_5O_3S$	$C_9H_6Cl_5O_4S$	
Form	Colorless crystal	Colorless crystal	-	
Melting point	109.2	213.3	181-201	
Vapor pressure	9.6×10^{-4} Pa (20°C)	4.0×10^{-5} Pa (20°C)	1×10^{-5} Pa (20°C)	
Octanol water partition coefficient	Log P_{OW} = 4.63-4.74	Log P_{OW} = 4.34-4.79	Log P_{OW} = 3.77	Sarafin and Asshauer 1987b; Muehlinger and Lemke 2004; Shen and Wania 2005
Henry's law constant	1.48 Pa m^3 mol^{-1} (25°C) 0.70 Pa m^3 mol^{-1} (25°C, FAV*) Temperature dependence: log H' = -876.14 / T + 0.4463 H (298K) = 7.95 (Pa m^3/ mol)	0.07 Pa m^3 mol^{-1} (25°C) 0.045 Pa m^3 mol^{-1} (25°C, FAV*)	8.46×10^{-3} Pa m^3 mol^{-1} (20°C, pH 5)	Weller (1990); Shen and Wania (2005); Rice et al., (1997)
Solubility	0.33 mg/L (22°C, pH 5) 0.53 mg/L (25°C, pH 5) 0.0063 mmol/L (25°C, FAV*)	0.32 mg/L (22°C, pH 5) 0.28 mg/L (25°C, pH 5) 0.089 mmol/L (25°C, FAV*)	0.50 mg/L (20°C, pH 5)	Sarafin and Asshauer (1987a); Goerlitz (1986); Weil et al. (1974); Shen and Wania (2005)
Hydrolysis DT_{50}	>1 yr (22°C, pH 5) 22 d (22°C, pH 7) 7 hrs (22°C, pH 9) 68 days (1°C, pH 8)	>1 yr (22°C, pH 5) 17 d (22°C, pH 7) 5 hrs (22°C, pH 9) 68 days (1°C, pH 8	-	Goerlitz and Kloeckner (1982); Sneikus (2005)
Photo-stability in water	Stable	Stable	-	Stumpf and Jordan (1993)

Endosulfan is a hydrophobic and non-polar compound. It is a non-combustible solid and slightly soluble in water, but can readily dissolve in organic solvents, for instance, xylene, chloroform, and kerosene. It can easily mix with common fungicides and other pesticides. The solubility of α- and β-isomers in water is reported to have 0.32 and 0.33 mgL^{-1}, respectively, at 20°C [19].

The melting point of technical endosulfan ranged from 70 -100°C [17] with a vapor pressure of 0.83 mPa at 20°C. This indicates an intermediate to the high volatility of endosulfan under field conditions [19]. The Henry's law constants of α- and β- endosulfan was determined to be 4.54×10^{-5} atmm^{3}mol^{-1} and 4.39×10^{-5} atmm^{3}mol^{-1}, respectively, indicating similar volatilization potential from water or moist soil surfaces [17]. The bioaccumulation potential of endosulfan in biota is (log Kow) 3.55 [17]. Endosulfan is a non-ionic compound, hence, subject to little or no dissociation at pH 5.0 to 9.0. The detailed physical and chemical characteristics of α-endosulfan, β-endosulfan, and endosulfan sulfate are listed in Table 2.

Technical endosulfan can be changed to numerous products in the environment with endosulfan sulfate as the predominant compound. The other byproduct of endosulfan includes endosulfan diol, endosulfan hydroxycarboxylic acid and endosulfan lactone [20]. The depuration rate of endosulfan residues in aquatic invertebrates and fish is faster than other organisms.

Toledo and Jonsson [21] opined the depuration half-lives of 2.9 for the α- endosulfan and 5.1 days for β-isomers. Likewise, the 5.9 days of depuration half-lives were estimated for the endosulfan sulfate transformation product in zebrafish (Brachydanio rerio). Ernst [22] discovered a depuration half-life of 34 hours for the α-isomer in marine mussels (*Mytilus edulis*).

3. WORLDWIDE PRODUCTION AND CONSUMPTION OF ENDOSULFAN

Endosulfan is most dominantly utilized in agriculture as an insecticide. It was first introduced as a broad spectrum insecticide by German scientist Farbwerke Hoechst in 1954 [23]. Later, endosulfan became an essential agrochemical and pest control agent globally to control a wide range of insect pests for a wide range of applications. Basically, limited data is accessible about the production of endosulfan. Worldwide, the annual production of endosulfan was determined to be 10,000 tonnes in 1984 [24]. However, the current production is believed to be significantly higher than what is estimated.

The Joint Canada-Philippines Planning Committee [25] estimated the worldwide the annual consumption of endosulfan for the last ten years with peaked production (approx. 13,000 tonnes) in 1998 and 2001. A recent finding by Li and MacDonald [26] is predictable with these consumption profiles. According to Li and MacDonald [26], the annual production of endosulfan is estimated to be around 12,800 tonnes (t), with India being the largest producer globally. It is estimated that India is producing about 5400 t yearly with its six plants [2]. The consumption rate of endosulfan in India was estimated to be about 113,000 t from 1958 to 2000. The United States was ranked second highest consuming country with 26,000 t from 1954 to 2000 [26]. Likewise, the annual consumption of endosulfan in China was estimated to be around 2800 t from 1998 to 2004 [27]. Contrarily, the consumption of endosulfan in European Union (based on sales data) showed 54% reduction from 1028 t/y (1995) to 469 t/y (1999) [2]. In total, the combined worldwide utilization of endosulfan in the agriculture sector from 1950 to 2000 is determined to be 308,000 t [26]. The worldwide consumption trend of endosulfan during 1996– 2004 is shown in Figure 1. Although consumption of endosulfan showed a declining trend in the northern hemisphere, the southern hemisphere has increased consumption rate (e.g., South America, Australia), keeping up yearly normal worldwide utilization of 12,450 t over the period 2000 to 2004 [28]. The gradual decline in global consumption of endosulfan as appeared in Figure 1 could be because of the presence of new insecticides (pyrethroids, neonicotinoids) in the market and also due to ongoing elimination of endosulfan by some country. Furthermore, the inclusion of endosulfan as POPs after the Stockholm Convention on POPs in 2011, resulted in a decline in worldwide consumption of endosulfan [11]. The usage of endosulfan is expected to diminish in the adopting countries over the next few years, while this consumption trend may continue in those countries that have not adopted this provision.

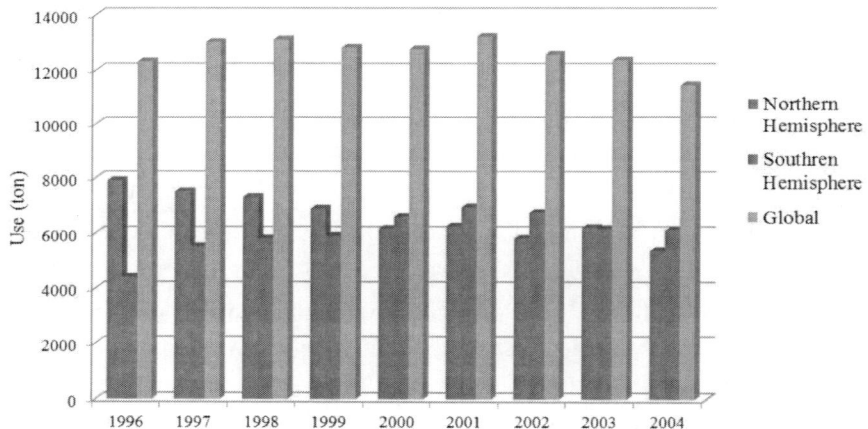

Figure 1. Worldwide consumption of endosulfan from 1996-2004 (Redrawn after Mackay and Arnold, 2005).

India is the largest producer and consumer of endosulfan in the world. Germany and the United States rank second and third in the world consumption list, respectively. Brazil, Australia, Sudan, and the former Soviet Union are the other important endosulfan consuming nation. In India, the production of endosulfan began in 1996 [29]. India had turned into the main producer of endosulfan by 2004, with a total production capacity of about 9,900 t [30]. India traded around 4,103.93 t of endosulfan to 31 countries in 2007-08 [31]. Coromandel Fertilisers Ltd, Excel Crop Care, and Hindustan Insecticides Ltd are the three major manufacturing units in India [32-33]. Excel crop care plant located in Bhavnagar in Gujarat is the biggest unit in India with a yearly production capacity of 6,000 t [34]. The Government of India owned Hindustan Insecticides limited located in Kerala which produces 1,600 tonnes of endosulfan annually [35].

Germany is the second most endosulfan producing country after India. The Bayer Crop Science located in Frankfurt produces approximately 4,000 t of endosulfan every year. Their major trading regions are Southeast Asia, Latin America, and the Caribbean [36]. Likewise, China produces about 2,400 t of endosulfan annually through three manufacturing units located in Jiangsu province, and 40 formulators mostly located in Shangdong and Jiangsu provinces [27, 37]. In 2007, four manufacturing units namely Kuaida Agrochemical Co, Jiangsu Xuzhou Shengnong Chemicals Co, Huangma Agrochemical Co, and Zhangjiagang Tianheng Chemical Co were in operation in Jiangsu [38]. Endosulfan is additionally produced in Israel (Makhteshim-Agan Industries) and is accounted for manufactured in South Korea.

4. EFFECTS OF ENDOSULFAN

4.1. Health Effects and Poisoning

The health effects of endosulfan depend on the amount of the endosulfan exposed and the duration of the exposure. It is opined that direct exposure of endosulfan or exposure through water, soil, or food to the public are generally lower than the level reported in the animal study. However, high levels of endosulfan may enter into the individual body either intentionally or unintentionally through contaminated food, during pesticide spray. The person exposed to high level of endosulfan suffered from tremors and seizures and may lead to death. Similar types of effect have been reported in an animal exposed to medium to high level of endosulfan.

Dietary intake is considered as the principal pathway of endosulfan exposure to human. Breathing of contaminated air or drinking of polluted water, smoking cigarettes from endosulfan contaminated tobacco, and working at endosulfan production center may also lead to exposure of high doses of endosulfan. Generally, a person working in the farm are exposed to high level of endosulfan than common man through direct handling and application or exposure through farm sprayed with endosulfan. Additionally, occupational exposure is another important medium of potentially high level of endosulfan exposure in the general population living in the vicinity of endosulfan production or disposal sites.

The toxicity of technical endosulfan is not much different from their metabolite or reaction product. However, the α-endosulfan is about 3 times more toxic than the β- endosulfan. Hyper-stimulation of the central nervous system and symptoms of seizures are the main toxicological impact of endosulfan. Other impacts include tremors, decreased respiration, dyspnea, salivation, tonic-clonic convulsions, and death in the latter case. Endosulfan can inhibit neurotransmitter system by antagonizing gamma-aminobutyric acid (GABA) function. A person with an accumulation of significant amount of endosulfan may also result in hepatic disorder, renal failure, and cardiac toxicities, hypotension, nausea, vomiting, confusion, tremors and coma. Respiratory failure and biochemical instability as a consequence of extensive seizure activity may lead to death within 48 h. Additionally, introduction to the high level of endosulfan has been known to cause mutations and alteration of DNA in human and other animals [39]. High doses of endosulfan in white clover and earthworm may cause genetic toxicity [40].

Numerous instances of intentional and unintentional poisonings have been reported from Asia, Africa and Latin America leading to death or severe disability. The effects of endosulfan poisoning include congenital deformities, late male sexual maturity, hormonal disorders in the female, congenital mental retardation, cerebral palsy, psychiatric disorder, epilepsy, cancers, skin, eye, ear, nose and throat problems, impaired memory, and chronic malaise. The main acute effect of endosulfan poisoning in survivors is hyperexcitation and convulsions, respiratory illness and heart problem. Endosulfan is an endocrine disruptor, oestrogenic and anti-androgenic in human cells, and causes breast cancer cells to grow. High doses of endosulfan may also interfere with male hormones, causing chronic depression of testosterone, suppresses the immune system, as well as promoting allergic responses. It is connected to long-term neurological impact and may cause Parkinson's disease. Birth defects have been also reported in experimental animal and human populations exposed to high level of endosulfan. Numerous examinations confirmed the mutagenic and genotoxic impact of endosulfan and finally leading to cancer in both experimental animals and human.

5.2. Effects on the Environment

Ecological pollution is far-reaching and has been detected in soil, groundwater and surface waters, marine sediments, air, rainfall, snow and ice pack, grasses and tree bark far away from endosulfan use area. Endosulfan is extremely lethal to wide ranges of organisms [36]. Aquatic ecosystems are more vulnerable to endosulfan contamination and may persist for a long duration in the environment leading to biomagnification in terrestrial food chains. It is very dangerous to aquatic organisms even at prescribed limits of application [41-43]. It is especially dangerous to fishes [42, 44, 45]. Massive killing of fishes has been reported [46]. It also disrupts the endocrine system, decrease of protein in tissues and other health impacts. Investigation on *Gambusia affinis* revealed high toxicity of endosulfan to fish species [45]. It also impacts the metabolic process in freshwater fishes by constraining transcription [47]. The pheromonal systems leading to disrupted male choice, reduction in female matting red-spotted newts has been also known due to endosulfan poisoning [48]. It also affects the germ cell population in the embryos of zebrafish [49].

Studies from South East Asia and Southern Pacific have demonstrated that endosulfan severely affects aquatic biota [50]. Additionally, high level of endosulfan is also lethal to frogs, toads, annelids, snails, aquatic insects, crustaceans, fishes and mollusks [51-53]. The hatching and survival rate of larval Indian Toad is also affected essentially due to endosulfan poisoning [54]. Studies from Argentina and South Africa revealed that endosulfan affects the aquatic insect and macro-invertebrate in dwelling in waterbodies [55-56].

A significant reduction in the growth rate and total protein content in earthworm may result from endosulfan poisoning [57]. Endosulfan is toxic to non-target organisms [58, 59], such as predators [60], and soil microarthropods [61], microorganisms, zooplankton [62], phytoplankton, soil algae, actinomycetes, and bacterial colonies etc [46, 63]. Studies have shown chromosomal aberrations in *Drosophila* fly [42], rabbits [46, 63] and rats [64, 65]. According to the National Wildlife Federation (USA), endosulfan is highly lethal to wildlife and intensively harmful to honey bees [46]. This is also hazardous to birds - mallard ducks [66], quails and pheasants [63].

Endosulfan has been known to cause phytotoxic effects on plants [67]. The effect of endosulfan on plants includes inhibition in root growth, stunting, burning of tips and edge of leaves and reduction in root permeability potential [46]. It is a conspicuous contaminant in vascular plants and lichen [67, 68]. Endosulfan is toxic to freshwater green algae [62] and blue-green algae [69]. Additionally, it affects the abundance of diatom, chrysophytes, cryptophytes, and dinoflagellates [70].

5. Environmental Fate of Endosulfan

Endosulfan is a semi-volatile organochlorine insecticide. Upon application in the field, it can evaporate from the soil surface and plants. Following dissipation, it can experience long-range atmospheric transport, thereby depositing out the air in colder place of the world. The fate of endosulfan released in the environment relies upon the medium it gets stored and is distinctive for α- and β-isomers. Endosulfan can be discharged into the air, water, and soil in areas where it is connected as a pesticide. The endosulfan sulfate is a reaction product found in technical endosulfan and can found in the environment due to oxidation or biotransformation. Once it discharged into the environment, endosulfan may undergo a variety of transformation and transport processes. This pesticide can travel a longer distance in the air before it finally ends up on crops, soil, and water. A number of studies have stated the transport of residue of endosulfan roughly

121 miles away from the agrarian sites of application [1, 71]. This insecticide may remain in the soil for a longer duration before its breakdown. Rainwater is another important agent which can wash endosulfan from soil to surface water. However, endosulfan may not readily dissolve in water and can stay in water attached to the soil [1].

The concentration of endosulfan in the air is highly variable relying upon the location. Remote areas generally have higher concentrations of endosulfan (18–82 pg/m^3), with peaked during crop seasons. After its release into the atmosphere, endosulfan will react with hydroxyl radicals with an estimated vapor phase half-life of 123 hr. Photolysis may not be an imperative process depending on the stability of thin films of endosulfan sulfate to light > 300 nm. In the air, α- and β-endosulfan may degrade chemically, but not by direct sunlight.

Generally, the contamination of endosulfan has not been reported in groundwater. However, the amounts of endosulfan in surface water are highly variable but are generally highest in water bodies receiving surface runoff from agricultural farm treated with endosulfan. Both the α- and β-endosulfan will be transformed into the less toxic endosulfan diol in waterbodies. Endosulfan sulfate is more difficult to break down in water.

Endosulfan is applied directly to plants and soil during its use as a pesticide. In soil, endosulfan attaches to soil particles and is not expected to move from soil to groundwater. Though α- and β-endosulfan can degrade in soil, but endosulfan sulfate is more resistant. Upon release of endosulfan sulfate in the soil, it will bind the soil, and prevent the leaching to the groundwater. Very little is known about the hydrolysis of endosulfan in soils. The hydrolysis half-lives of α- and β-endosulfan are 35 to 37 days at pH 7.0 and 150 to 187 days at pH 5.5, respectively. Biological degradation in soil with a half-life of 11 weeks is another important fate process for endosulfan sulfate.

The transfer of endosulfan isomers from soil to air is an important step in understanding the fate of endosulfan. Foods, especially crisp and solidified vegetables are the significant medium which accumulates a significant amount of endosulfan. Upon consumption by mammals, endosulfan gets rapidly breakdown and removed with a little absorption in the gastrointestinal tract. The endosulfan sulfate has shown similar acute toxicity to the parent compound. The β- isomer gets clear from blood plasma more quickly than the α-isomer.

6. REGULATORY STATUS OF ENDOSULFAN UNDER INTERNATIONAL CONVENTION

Despite endosulfan has been classified as a POPs, yet was excluded in the previous list focusing for elimination under the Stockholm Convention. However, this was recorded as a POPs in the convention on LRTAP. United Nation Environment Program (UNEP) has also recognized endosulfan as PTS. Endosulfan has been in overall use since its presentation in the 1950s and was known to a safer replacement to other OCPs until the 1970s. In the previous two decades, the majorities of the countries have categorized endosulfan as a hazardous substance and have forbidden or regulated its usage. Nations which have prohibited the usage of endosulfan comprised Singapore, Belize, Tonga, Syria, Germany, Sweden, Philippines, Netherlands, St. Lucia, Columbia, Cambodia, Bahrain, Kuwait, Oman, Qatar, Saudi Arabia, United Arab Emirates, Sri Lanka and Pakistan [41, 50]. Limited use of endosulfan is permitted in Australia, Bangladesh, Indonesia, Iran, Japan, Korea, Kazakhstan, Lithuania, Thailand, Taiwan, Denmark, Serbia & Montenegro, Norway, Finland, Russia, Venezuela, Dominican Republic, Honduras, Panama, Iceland, Canada, the United States and the United Kingdom [24, 41, 50, 67].

In 2007, the European Union Council proposed to include the endosulfan in the global POPs elimination list under the Stockholm Convention on POPs [72]. Later, in 2008, the Persistent Organic Pollutants Review Committee accepted the proposal of including endosulfan in POPs list [73], and was accepted in the fifth meeting held in 2009. Likewise, the Chemical Review Committee of the Rotterdam Convention on the Prior Informed Consent Procedure agreed to recommend endosulfan to the Conference of the Parties of the Convention for including in the list of chemicals that have been banned or severely restricted for health or environmental reasons.

Endosulfan has been recorded as the priority pollutants in Annex X of the European Union's water policy's [74]. It is also incorporated in the OSPAR's priority chemical action list for the Protection of the Marine Environment in the North-East Atlantic [75]. The Third North Sea Conference has also listed endosulfan as the priority substances. The Regional-based Assessments of Persistent Toxic Substances by UNEP-GEF (PTS) included Indian Ocean region—regional concern [29]; North American region—regionally specific PTS [76]; Mediterranean region—local concern [77]; Sub Saharan Africa—PTS of highest concern after DDT [78]; East and West South American—emerging concern [79]; European region—proposed possible priority hazardous substance [80]; South East Asia and South Pacific region—regional concern, with long-term effect on the structure of the aquatic ecosystem [81]; and Central America and Caribbean—one of the most important PTS of emerging concern [82].

CONCLUSION

In spite of the fact that endosulfan has been prohibited or limited in various nations, it is still being widely utilized in different nations. Endosulfan is intensely poisonous and has been implicated in many cases of poisoning and fatalities. Additionally, it is profoundly lethal to the environment. It has been related to a range of chronic effects similar to organochlorine pesticide, for instances, malignancy and effects on the hormonal disorder in individual exposed to endosulfan. The rapid dissemination of the endosulfan isomers is identified with instability and it is then subject to atmospheric long-range transport. A combination of experimental data, models, and monitoring results have confirmed the persistence, specifically in a colder region, bioaccumulation potential and LRAT characteristics of endosulfan. In view of the inherent properties, and ubiquitous presence and biota in remote territories, it is inferred that endosulfan is likely, because of its long-range atmospheric transport. Hence, global action is warranted to prevent significant health and environmental effects.

REFERENCES

[1] 1ATSDR, 2000. *Toxicological profile for endosulfan*. US Department of Health and Human Services. Public Health Service. Available at:,http://www.atsdr.cdc.gov/Tox Profiles/ tp41.pdf.

[2] Ayres, R. U., Ayres, L. W., 2000. The life cycle of chlorine, Part IV: Accounting for persistent cyclic organochlorines. *J Ind Ecol* 4, 121–59.

[3] Botello, A. V., Rueda-Quintana, L., Díaz-González, G., Toledo, A., 2000. Persistent organochlorine pesticides (POPs) in coastal lagoons of the subtropical Mexican Pacific. *Bull Environ Contam Toxicol* 64, 390–7.

[4] Herrmann, M., 2000. *Preliminary risk profile of endosulfan*. Berlin Germany: Umwelt bundesamt.

[5] Laabs, V., Amelung, W., Fent, G., Zech, W., Kubiak R. 2002. Fate of 14C-labeled soybean and corn pesticides in tropical soils of Brazil under laboratory conditions. *J Agric Food Chem* 50, 4619–27.

[6] Laabs, V., Amelung, W., Pinto, A. A., Wantzen, M., da Silva C. J., Zech, W., 2002. Pesticides in surface water, sediment, and rainfall of the northeastern Pantanal basin, Brazil. *J Environ Qual* 31, 1636–48.

[7] OSPAR.2002. OSPAR—Background Document on Endosulphan. Hazardous Substance Series. OSPAR Commission. Oslo: Oslo–Paris Convention for the North–East Atlantic0946956987; p. 1–42.

[8] Shen, L., Wania, F., Ying, D. L, Teixeira, C., Muir, D. C. G., Bidleman, T. F., 2005. Atmospheric distribution and long-range transport behaviour of organochlorine pesticides in North America. *Environ Sci Technol* 39:409–20.

[9] Roberts, D. M, Karunarathna, A., Buckley, N. A, Manuweera, G., Sheriff, M. H. R., Eddleston, M. 2003. Influence of pesticide

regulation on acute poisoning deaths in Sri Lanka. *Bull World Health Organ* 8:789–98.

[10] US-EPA. 2007. *RED (Re-registration Eligibility Decision) document: endosulfan updated risk assessments, notice of availability, and solicitation of usage information.* Federal Register United States Environmental Protection Agency. p. 64624–6. Docket: EPA-HQ-OPP-2002-0262. (July 2007).

[11] FAO. 2011. *FAO specifications and evaluations for agricultural pesticides. Endosulfan (1,4,5,6,7,7- hexachloro-8,9,10-trinorborn-5-en-2,3-ylenebismethylene)sulfite.* Food and Agriculture Organization of the United Nations, 1-17.

[12] Muller, F., Streibert, H. P., Farooq, M., 2009. Acaricides. In: Ullmann's encyclopedia of industrial chemistry. Vol. 1. Weinheim, Germany: Wiley-VCH Verlag GmbH & Co., 91-190.

[13] Capkin, E., Altinok, I., Karahan, S., 2006. Water quality and fish size affect toxicity of endosulfan, an organochlorine pesticide to rainbow trout. *Chemosphere* 64 (10), 1793–1800.

[14] Ravi, K., Varma, M. N., 1998. Biochemical studies on endosulfan toxicity in different age groups of rats. *Toxicol. Lett.* 44 (3), 247–252.

[15] Bharath, B. K., Srilatha, C., Anjaneyulu, Y., 2011. Reproductive toxicity of endosulfan on male rats. *Int. J. Pharma Bio Sci.* 2 (3), 508–512.

[16] IFCS. 2003. *Acutely Toxic Pesticides: Initial Input on Extent of Problem and Guidance for Risk Management. Forum Standing Committee Working Group.* Forum IV Fourth Session of the Intergovernmental Forum on Chemical Safety, Bangkok, Thailand, Nov 1-7.

[17] Mackay, D., Shiu, W. Y., Ma, K. C., 1997. Illustrated Handbook of Physical-Chemical Properties and Environmental Fate for Organic Chemicals. Vol.5. Pesticide chemicals. Lewis Publishers, New York. 812 pp.

[18] Dureja, P., Mukerjee, S. K., 1982. Photoinduced reactions: Part IV. Studies on the photochemical fate of 6,7,8,9,10,10-hexachloro-1,5,5a,6,9,9a-hexahydro-6,9-methano-2,4,3-benzo(e)dioxathiepin-3-oxide (endosulfan), an important insecticide. *Indian J Chem* 21B:411-413.
[19] Tomlin, C. D. S. 2000. *The pesticide manual: A world compendium.* The British Crop Protection Council, Surrey, UK.
[20] German Federal Environment Agency-Umweltbundesamy, Berlin. 2004. Endosulfan. Draft Dossier. 64 pp.
[21] Toledo, M. C. F., Jonsson, C. M., 1992. Bioaccumulation and elimination of endosulfan in zebra fish (Brachydanio rerio). *Pesticide Science* 36:207-211.
[22] Ernst, W. 1977. Determination of the bioconcentration potential of marine organisms - a steady state approach. I. Bioconcentration data for seven chlorinated pesticides in mussels (Mytilus edulis) and their relation to solubility data. *Chemosphere,* 6:731-740.
[23] Maier-Bode, H., 1968. Properties, effect, residues and analytics of the insecticide endosulfan. *Resid. Rev.* 22:1-44.
[24] Anon, 2003. Pesticide News No 60, *The Journal of Pesticide Action Network UK.* (Quarterly), P 19.
[25] Joint Canada-Philippines Planning Committee. 1995. International Experts Meeting on Persistent Organic Pollutants Towards Global Action, The Joint Canada-Philippines Committee Volumes I-IV.
[26] Li, Y. F., Macdonald, R. W., 2005. Sources and pathways of selected organochlorine pesticides to the Arctic and the effect of pathway divergence on HCH trends in biota: a review. *Sci. Tot. Environ.,* 342: 87-106.
[27] Jia H, Li Y. F, Wang D, Cai D, Yang M, Ma J., 2009. Endosulfan in China 1: gridded usage inventories. *Environ Sci Pollut Res* 16:295–301.
[28] Mackay, N., Arnold, D. 2005. Evaluation and interpretation of environmental data on endosulfan in Arctic regions. Report for Bayer CropScience, Cambridge Environmental Assessments (CEA), Report number 107. UK: Cambridge; 2005.

[29] GEF IO. 2002. Regionally Based Assessment of Persistent Toxic Substances – Indian Ocean Regional Report. Global Environment Facility, United Nations Environmental Programme, Geneva. http://www.chem.unep.ch/Pts/.

[30] MCF. 2008. Performance of Chemical & Petrochemical Industry at a Glance (2001-2007). Monitoring and Evaluation Division, Department of Chemicals & Petrochemicals, Ministry of Chemicals & Fertilizers, Government of India, New Delhi. http://www.chemicals.nic.in/stat0107. pdf.

[31] MCI. 2008. Export Import Data Bank. Export: Commodity-wise all countries. Commodity 38081018. Endosulfan technical. Department of Commerce, Ministry of Commerce & Industry, Government of India. http://commerce.nic.in/eidb/Default.asp.

[32] Venkatraman, L., 2004. Will banning endosulfan solve the problem? The Hindu Business Line. July 6. http://www.thehindubusinessline.com/2004/07/06/stories /2004070600790300.htm.

[33] CFL. 2008. Coromandel Fertilsers Ltd. Profile. http://www.murugappa.com/companies/coromandel/overview.htm#pro.

[34] Yadav, K. P. S., Jeevan, S. S., 2001. Endosulfan conspiracy. Centre for Science and Environment, New Delhi https://www.downtoearth.org.in/coverage/endosulfan-conspiracy-38732

[35] HIL. 2008. Hindustan Insecticides Ltd Manufacturing Units. http://www. hil-india.com/manufacturingunits.htm.

[36] GFEA-U. 2007. Endosulfan. Draft Dossier prepared in support of a proposal of endosulfan to be considered as a candidate for inclusion in the CLRTAP protocol on persistent organic pollutants. German Federal Environment Agency – Umweltbundesamt, Berlin.

[37] Li J., Zhang, G., Guo L, Xu W, Li X, Lee, C.S.L, Ding, A., Wan, T., 2007. Organochlorine pesticides in the atmosphere of Guangzhou and Hong Kong: Regional sources and long-range atmospheric transport. *Atmos Environ* 41:3889-903.

[38] Sun Jing. 2007. Email communication, July 27. Pesticide Eco Alternatives, China.

[39] Bajpayee, M., Pandey, A. K., Zaidi, S., Musarrat, J., Parmar, D., Mathur, N., 2006. DNA damage and mutagenicity induced by endosulfan and its metabolites. *Environmental and Molecular Mutagenesis,* 47(9), 682-692.

[40] Liu, W., Zhu, L. S., Wang, J., Wang, J. H., Xie, H., & Song, Y., 2009. Assessment of the genotoxicity of endosulfan in earthworm and white clover plants using the comet assay. *Archives of Environmental Contamination and Toxicology,* 56(4), 742746.

[41] Anon, 2002. Regional Based Assessment of Persistent Toxic Substances – Indian Ocean Regional Report – Chemicals – United Nations Environmental Programme – Global Environment Facility (UNEP – GEF)

[42] Michael, H., 2003. Endosulfan Preliminary Dossier; www.unece.org/env /popsxg/docs/2000-2003/dossier-endosulfan-may03.pdf.

[43] Leonard, A. W., Hyna, R. V., Leigh, K. A, Le. J, Beckett, R., 2001. Fate and toxicity of endosulfan in Naomi River water and bottom sediments. *J. Environ. Qual.* 2001May-June; 30(3):750-9.

[44] Anon, 1984. Environment Health Criteria 40- Endosulfan. IPCS (International Programme on Chemical Safety) – WHO Geneva

[45] Karim, A. A, Haridi A. A, Rayah, E A., 1985. The environmental impacts of four insecticides on non-target organisms in the Gezira Irrigation Scheme canals of Sudan. *J. Trop. Medic. Hyg:* Vol 88, ISS 2, P 161-8.

[46] Romeo, F., Quijano, M. D., 2000. Risk Assessment in a third world reality: An Endosulfan case History. *International Journal of Occupational and Environment Health.* Vol. 6, No. 4.

[47] Tripathi, G., Verma, P., 2004. Endosulfan mediated biochemical changes in the fresh water fish Clarias batrachus. *Biomed Environ Sci* 17 (1) 47-56.

[48] Park, D., Hempleman, S. C., Propper, C. R., 2001. Endosulfan exposure disrupts Pheromonal system in the red spotted Newt- A Mechanism for subtle effects of environmental chemicals. *Environmental Health Perspectives* 109(7); 669-673.

[49] Willey, J. B., Krone, P. H., 2001. Effect of endosulfan and nonyl phenol on the primordial germ cell population in prenatal zebrafish embryos. *Aquat. Toxicol.* 54(1-2); 113-123.

[50] Anon, 2002. Regional Based Assessment of Persistent Toxic Substances- Central America and the Caribbean- Regional Report – Chemicals- United Nations Environmental Programme Global Environment Facility.

[51] Anon, 2001. PAN Pesticide Database – Aquatic Ecotoxicity Studies for Endosulfan. (Derived from USEPA AQUIRE Acute Summaries).Pesticide Action Network, Penang, Malaysia.

[52] Goulet, B. N., Hastela, A., 2003. Toxicity of cadmium, endosulfan and atrazine in adrenal strereoidogenic cells of two amphibian species, Xenopus laevii and Ras catesbeiana *Environ Toxicol Chem* 2003 Sep;22(9)2;106-13.

[53] Otludil B., Cengiz E. I., Yildirim M. Z., Unver O., Unlu, E., 2004. The effects of endosulfan on the great ramshorn snail Planorbarius corneus (gastropoda,pulmonata) a histopathological study. *Chemosphere* 56 (7): 707-16.

[54] Mercy, Mathew, Andrews, M. I., 2000. Effect of Endosulfan on the hatching rate and larval survival of Common Indian Toad (Bufo melanostictus Schneider), Proceedings of the 12th Kerala Science Congress, Government of Kerala, pp 615-17.

[55] Jergentz, S., Mugin, H., Bonetto, C, Schulz, R., 2004. Runoff related endosulfan contaminations and aquatic macro invertebrate responses in rural basins near Buenos Aires, Argentina. *Arch. Environ Contam.Toxicol.* 46 (3):345-352.

[56] Thiere, G., Schuz, R., 2004. Runoff related agricultural impact in relation to macro invertebrate communities of the Lourens river, South Africa. *Water Res* 38(13) 3092-102.

[57] Mosleh, Y. Y., Paris-Palacois, S., Couderchet, M., Vernet, G., 2003. Acute and sub-lethal effects of two insecticides on earthworm (Lumbricus terrestris L) under laboratory conditions. *Environ Toxicol.* 18 (1); 1-8.

[58] Park, E. K., Lees, E. M., 2004. The interaction of endosulfan with the collembolan Proistoma minuta (Tullberg): toxicity, the effects of sublethal concentrations and metabolism. *Pest manag. Sci.* 60(7), 710-8.

[59] Naqvi, S. M., Vaishnavi, C., 1993. Bio Accumulative Potential and toxicity of Endosulfan insecticide to non-target animals. *Comp. Biochem. Physiol. C*; Vol 105, Iss 3, P 347-61.

[60] Elizen, G. W., 2001. Lethal and Sublethal effects of insecticide residues on Orius insidiorus (Hemiptera, Anthocoridae) and Geocoris punctipes (Hemiptera, Lygaeidae) *Journal of Economic Entomology* 94(1); 55-59.

[61] Joy, V. C., Chakravarthy, P. P., 1991. Impact of Insecticide on Non Target Micro Arthropod Fauna in Agricultural Soil. *Ecotoxicol. Environ. Saf.;* Vol. 22, ISS 1, P 8-16.

[62] Delorenzo, M. E., Taylor, L. A., Lund, S. A., Pennington, P. L., Strozier, E. D., Fulton, M. H., 2002. Toxicity and bio-concentration potential of agricultural pesticide endosulfan in phytoplankton and zooplankton. *Arch. Envron. Contam. Toxicol.* 42(2):173-81.

[63] Susan, S., Sania, P., 1999. Endosulfan- A Review of its Toxicity and its Effects on the Endocrine System WWF (World Wild Life Fund – Canada).

[64] Kalendar, S., Kalendar, Y., Ogutcu, A., Uzunhisarcikli, M., Durak, D., Acikgoz, F., 2004. Endosulfan-induced cardio-toxicity and free radical metabolism in rats.the protective effect of Vitamin E. *Toxicology* 202(3); 227-35.

[65] Reuber, M. D., 1981. The role of toxicity in the carcinogenicity of Endosulfan. *Sci. Total Environ.* 20(1); 23-47.

[66] Anon, 1996. Pesticide information Profile- Endosulfan (Revised June 1996) EXTENT- Extention Toxicology Network.

[67] Anon, 2002. End of the Road for Endosulfan- A Call for Action Against A Dangerous Pesticide Environmental Justice Foundation, London, UK. Internet site- www.ejfoundation.org.

[68] Vorkamp, K., Riget, F., Glasius, M., Pecseli, M., Lebeuf, M., Muir, D., 2004. Chlorobenzene, chlorinated pesticide coplanar

chlorobiphenyl,and other organochlorine compounds in Greenland Giota. *Sci Total Environ* 2004 Sep 20; 331(1-3);1.

[69] Tandon, R. S., L. S. Lal, Narayana, Rao.V.V., 1988. Interaction Endosulfan and Malathion with the green algae Anabaena and Autosira fertilissima. *Environ polute* 52(1); 1-9.

[70] Downing, H. F., De Lorenzo, M. E., Fulton, M. H., Scott, G. I., Madden, C. J., Kucklick, J. R., 2004. Effects of the agricultural pesticides atrazine, chlorothalonil and endosulfan on South florida microbial assemblages. *Ecotoxicology* 2004 Apr; 13 (3): 245-60.

[71] Bradford, D. F., Heithmar, E. M., Tallent-Halsell, N. G., Momplaisir, G. M., Rosal, C. G. Varner, K. E., 2010. Temporal patterns and sources of atmospherically deposite pesticides in alpine lakes of the Sierra Nevada, California, U.S.A. *Environmental Science & Technology,* 44(12), 46094614.

[72] EU. 2007. Press Release 11911/07 (Presse 170). 2816th Council Meeting, General Affairs. Council of the European Union, Brussels July 23.

[73] UNEP. 2008. POPRC-4/5: Endosulfan. http://chm.pops.int/Convention/ POPs Review Committee/ Meetings/POPRC4/ POPRC4Report and Decisions /tabid/450/ language/en-US/Default.aspx.

[74] EU. 2001. Decision No 2455/2001/EC of the European Parliament and of the Council of 20 November 2001, amending Directive 2000/60/EC. *Official Journal of the European Communities* L331/1. Dec 15. http://eur-lex.europa.eu/LexUriServ/ site/en/oj/2001/l_331/l_33120011215en00010005.pdf.

[75] OSPAR. 2006. OSPAR Convention For The Protection Of The Marine Environment Of The North East Atlantic. OSPAR List of Chemicals for Priority Action (Update 2006). (Reference number 2004-12). http://www. ospar.org/eng/html/welcome.html.

[76] GEF NA. 2002. Regionally Based Assessment of Persistent Toxic Substances – North America Regional Report. Global Environment Facility, United Nations Environmental Programme, Geneva. http://www.chem.unep.ch/Pts/.

[77] GEF M. 2002. Regionally Based Assessment of Persistent Toxic Substances – Mediterranean Regional Report. Global Environment Facility, United Nations Environmental Programme, Geneva. http://www.chem.unep.ch/Pts/.

[78] GEF SSA. 2002. Regionally Based Assessment of Persistent Toxic Substances – Sub-Saharan Africa Regional Report. Global Environment Facility, United Nations Environmental Programme, Geneva. http://www. chem.unep.ch/Pts/.

[79] GEF SA. 2002. Regionally Based Assessment of Persistent Toxic Substances – Eastern and Western South America Regional Report. Global Environment Facility, United Nations Environmental Programme, Geneva. http://www.chem.unep.ch/Pts/.

[80] GEF E. 2002. Regionally Based Assessment of Persistent Toxic Substances – Europe Regional Report. Global Environment Facility, United Nations Environmental Programme, Geneva. http://www.chem. unep.ch/Pts/.

[81] GEF SEA SP. 2002. Regionally Based Assessment of Persistent Toxic Substances – South East Asia and South Pacific Regional Report. Global Environment Facility, United Nations Environmental Programme, Geneva. http://www.chem.unep.ch/Pts/.

[82] GEF CAC. 2002. Regionally Based Assessment of Persistent Toxic Substances – Central America and the Caribbean Regional Report. Global Environment Facility, United Nations Environmental Programme, Geneva. http://www.chem.unep.ch/Pts/.

Chapter 2

ENDOSULFAN CONTAMINATION IN INDIAN ENVIRONMENT

Amrendra Kumar
and Ningombam Linthoingambi Devi, PhD*
Department of Environmental Science,
Central University of South Bihar, Bihar, India

ABSTRACT

Endosulfan is wide spectrum semi-volatile insecticide and one of the highly toxic organochlorine pesticides. It is a cyclic sulfite ester, i.e., 1, 5, 5a, 6, 9, 9a-hexahydro-6, 9-methano-2, 4, 3-benzodioxathiepine 3-oxide substituted by chloro groups at positions 6, 7, 8, 9, 10 and 10. Persistent, toxicity, bioaccumulation and atmospheric long-range transport are the important characteristic of the endosulfan governing its fate in the environment. Exposure to endosulfan can cause both acute as well as chronic toxicity in human. The adverse impact of endosulfan on environment component has been also reported worldwide. With realizing the adverse effects of endosulfan on human health and the environment, it was regulated in the fifth meeting of the Stockholm convention on

---

* Corresponding Author's Email: nldevi@cub.ac.in.

persistent organic pollutants (POPs) in 2011. Endosulfan and its metabolites have been broadly disseminated and their trace has been reported in multi-environment components. Soil and sediments component in Indian environment are the most contaminated matrices in India. In this comprehensive review, we evaluate and update the basics of endosulfan, its fate, and distribution in multi-environmental matrices. Additionally, the residual levels of endosulfan in India environment is reviewed and documented.

Keywords: endosulfan, persistent, bioaccumulation, atmosphere, India

1. INTRODUCTION

Endosulfan is one of the most prevalent organochlorine persistent organic pollutants (POPs) and has been reported globally including the Arctic region due to its persistent, toxic, semi-volatile and long-range transport nature [1-2]. Though endosulfan was first manufactured in the 1950s by Farbwerke Hoechst in the USA, it was registered in 1954 as a pesticide. It was established as a potential and effective chemical against a wide spectrum of insects and mites in agriculture and another related field [3-5]. Thus, endosulfan is applied in a wide range of crops. However, because of its ubiquitous presence, it can also contaminate different matrices of the environment, for instance, air, soil, water and vegetation [6 -10]. Endosulfan is a wide spectrum insecticide of the cyclodiene sub-group which is composed of two biologically active isomers i.e., α- and β-endosulfan in the ratio of 70:30 [11]. Two important metabolites i.e., endosulfan sulfate and cyclodiene are yet being used in various application around the world and has been reported in sediments, water, and animal tissues [4, 12]. The Stockholm Convention on POPs and the United Nations Environmental Protection (UNEP) has obliged all the signatory countries to reduce/restrict and eliminate the production and usages of endosulfan. They also categorized endosulfan as a dirty dozen compounds due to its toxicity properties.

India is the world's largest producer and user of endosulfan in the agricultural sector. During the year 1995-2000, the total production of endosulfan was determined to be around 41,033 tons with a yearly production of 8,206 tons/year. A number of studies have reported the contamination of endosulfan in different sectors of the environment including soil, water and sediments [13-23]. However, a limited study has been made dealing with the fate and distribution of endosulfan in Indian air compared with other environmental matrices. This chapter reviews and updates the knowledge and understanding the fundamentals of endosulfan and its residual contamination in Indian environment based on available literature.

2. STRUCTURE AND PROPERTIES OF ENDOSULFAN

The chemical structure of endosulfan and its metabolite has been shown in Figure 1. Endosulfan is a cyclic sulfite ester, i.e., 1, 5, 5a, 6, 9, 9a-hexahydro-6, 9-methano-2, 4, 3-benzodioxathiepine 3-oxide substituted by chloro groups at positions 6, 7, 8, 9, 10 and 10. It is a cyclodiene organochlorine insecticide and a cyclic sulfite ester. The endosulfan compounds are semi-volatile and are more vulnerable to evaporation to the air and get back deposited to the soil, water or vegetation [24-27]. The β-endosulfan has a slightly high melting point (208-210°C) than that with α-endosulfan (108-110°C) and endosulfan sulfate (181-201°C)(Table 1). The water solubility of α- and β- endosulfan are identical at 25°C, while the endosulfan sulfate metabolite has low water solubility [28]. Likewise, α- and β- endosulfan have similar vapor pressure while lower for endosulfan sulfate. The bioaccumulation potential (log Kow) of endosulfan isomers close to 5 indicates likely bioaccumulation in the environment [29]. The organic carbon partitioning coefficient (log Koc) of β-endosulfan is higher in soil and sediment than α-endosulfan [28, 31, 32]. The half-life of α-endosulfan is lower in soil than β-endosulfan and endosulfan sulfate under aerobic process [33]. Therefore, endosulfan isomers show their bioaccumulation potential towards the environment.

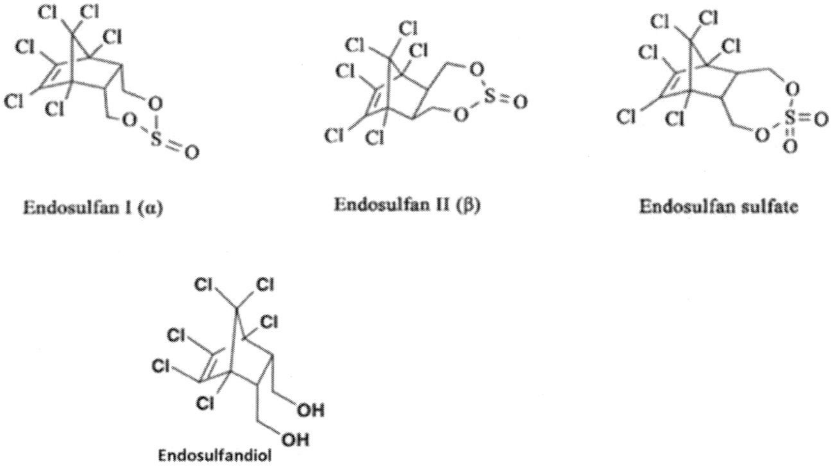

Figure1. Chemical structure of endosulfan and its metabolite [30].

The transport of endosulfan in the air is governed by its half-life period, and the ability to get volatilized from soil and plants. The rate of endosulfan evaporation increases with temperature, humidity, and wind [34]. Additionally, endosulfan in the air is also dependent on its interchange potential between air and water, and air and air-borne particles. These interchange mechanisms may likely to affect uptake of endosulfan into the atmosphere, transport to a remote place and its subsequent deposition. Endosulfan may also evaporate from water because the volatilization half-life from surface waters is greater than 11 days and possibly greater than 1 year [35]. Like POPs, endosulfan has a tendency to travel a longer distance from its source of origin and get deposited in remote environments at higher latitudes and elevations. This phenomenon is governed by low temperatures and efficient scavenging from the atmosphere [36, 37]. The α- endosulfan was first detected in 1986 in snow samples in the Canadian Arctic (Ellesmere Island) and the concentration ranged from non-detectable to 1.34 ng/m^3 [38, 39]. Later, during 2002-03, the residual concentration of endosulfan was regularly measured in snowpack samples from seven national parks in arctic, sub-arctic and alpine regions of the US. The Sequoia National Park measured the highest concentration of endosulfan with 1500 ng/m^3 and was said to be originated from regional transport. Residues level of endosulfan in the Arctic region measured with a pick concentration of 170 ng/m^3 [70]. Low concentration of Endosulfan and its metabolite have also been detected in marine waters and sediments [40].

3. FATE AND DISTRIBUTION OF ENDOSULFAN

The extensive use of endosulfan by human causes detection of endosulfan residue in different segments of the environment. The atmospheric contribution of endosulfan is primarily influenced by volatilization which depends on temperature and humidity of the environment and partitioning coefficient of the compound. Likewise, the process of volatilization from soil to the air is governed by sorption potential of the soil i.e., organic carbon content in the soil. The α-endosulfan partitioned easily in the atmosphere compare to β-endosulfan. It is opined that β-endosulfan can be converted into α-endosulfan in solid water interface and air-water interface.

After the release of endosulfan from the source site, it can directly or indirectly affect the atmosphere, vegetation, groundwater, and soil. In the atmosphere, endosulfan may travel a longer distance before it gets deposited on crops, soil, or water bodies. Vegetation gets affected due to the metabolism of endosulfan. Soil may also get affected due to direct absorption, run-off, and litterfall from the vegetation contaminated with endosulfan. Leaching of endosulfan through endosulfan contaminated soil and vegetation may directly affect groundwater, which in turn impacts the soil, crops and flora and fauna (Figure 2). From soil, the endosulfan may transfer to biotic species through plant uptake. Atmospheric washout by rain leads to deposition of endosulfan in water bodies, soil, and vegetation. The traces of endosulfan of waterbodies may bio-accumulate to aquatic flora and fauna, for instance, fish and can be biomagnified to the human beings through the food chain (Figure 3).

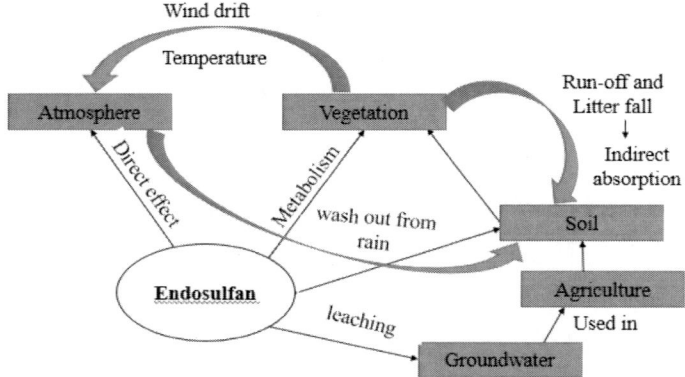

Figure 2. Distribution mechanism of endosulfan in the environment [70].

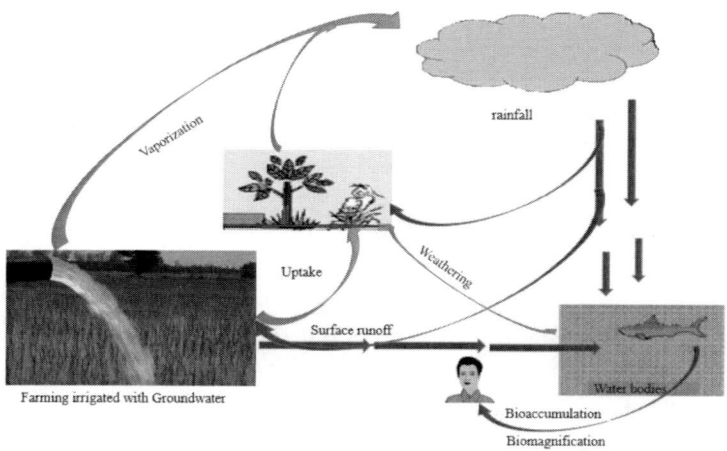

Figure 3. Distribution of endosulfan in groundwater environment [70].

4. Effects of Endosulfan

4.1. Health Impact

Endosulfan is highly toxic contaminants via ingestion, inhalation, and, dermal contact routes of exposure and it is associated to human poisoning [44, 45]. Exposure to endosulfan can cause acute as well as chronic toxicity in human [46, 47]. The primary effect of endosulfan, via oral and dermal routes of exposure, is on the central nervous system (CNS). Prolonged exposure of endosulfan is reported to cause mental retardation, loss of memory, congenital birth defects immunosuppression, chromosomal aberration, neurological disorders and dullness [48]. Exposure of endosulfan has been linked to congenital physical disorders, mental retardations and deaths in farm workers and villagers in developing countries in Africa, Asia and Latin America [49-52]. A survey conducted by PAN Africa in Mali in 2001 of villages in 21 areas of Kita, Fana, and Koutiala found a total 73 cases of pesticide poisoning and endosulfan was the main pesticide identified [52]. Endosulfan can also affect the function of enzymes, for instance, lactic dehydrogenase, glucose-6-phosphate dehydrogenase and alkaline phosphatase [53, 54]. Low level of endosulfan can also adversely affect the immune system [55]. Chronically, endosulfan is neurotoxic to the central nervous system [48, 55] tissue cultures [56, 57] and can inhibit GABA-A receptors [58, 59]. It can also cause genotoxic impact in human lymphocytes and DNA damage in Chinese hamster ovary (CHO) cells and mutation [60].

Endosulfan is also a famous endocrine disrupter and can affect the reproductive system as well as the kidney and liver etc. Exposure to endosulfan has been known to advocate tumor, cancer, and teratogen. High doses of endosulfan instantly may cause poisoning and even death in humans. The poisoning of endosulfan in humans leads to the lowering of blood pressure and electrocardiogram aberration [61]. Some patient can also develop symptoms of rhabdomyolysis, hepatic toxicity and hypotension [62]. Exposure to as low as 500 mg kg^{-1} of endosulfan has been reported to cause death in humans and, in many cases, led to permanent brain damage. Rashes and skin irritation have been reported due to chronic exposure of endosulfan in farm worker. The US Environmental Protection Agency has categorized endosulfan as class I (Highly Acutely Toxic) while the World Health Organization categorizes it as Class II (Moderately Hazardous) substance.

4.2. Impacts on the Environment

Endosulfan is a universal environmental contaminant and resistant to degradation. It can travel long distance far away from its source of generation. The impact of endosulfan on the different environmental component is discussed under following.

4.2.1. Impacts on Plants

Different types of chlorinated hydrocarbon insecticides including endosulfan have been tested to cause harmful effects on germination and growth of plants. It also inhibits the nitrification and mineralization processes. The application of endosulfan may also influence the macro and microelement constituents of corn and beans growing in soil. Both decrease and increase in uptake of the nutrient have been reported due to the high level of endosulfan.

4.2.2. Impacts on Aquatic Species

Endosulfan is equally toxic to fish and other biotas as human. It can disrupt the aquatic food chain and particularly toxic to juveniles [63]. It can cause a reduction in calcium and magnesium level of blood, inhibits brain ATP, inhibit acetylcholinesterase in the brain of fish and can cause massive fish kills [64]. Recently the genotoxic impact of endosulfan has been reported to freshwater fish leading to break of single-cell DNA strand breaks in the tissue of the gill, kidney, and erythrocytes [65]. The harmful effects of endosulfan have been also reported to Pacific oysters [66]. The binding of endosulfan to fish may alter the structure of DNA and increase the hyperchromicity [67]. The developmental and reproductive impact of endosulfan in non-target animals has been also reported [68]. Endosulfan is highly bioaccumulative in fish [69]. It is also toxic to amphibians. A small amount of endosulfan has been observed to kill tadpoles [70]. Exposure of level of endosulfan in aquatic invertebrates causes decreases in adenylate energy charge, oxygen consumption, hemolymph amino acids, succinate dehydrogenase, heartbeat (in mussels), and altered osmoregulation [71]. It

4.2.3. Impacts on Birds and Reptiles

Endosulfan is highly toxic to birds and reptiles. Endosulfan can suppress bird immune system [73]. Exposure to extremely low doses of endosulfan in chicken eggs can result in adverse effects on the liver and brain enzymes, decreased DNA and RNA in the brain, and immunosuppression [74]. Exposure to sublethal doses of endosulfan in chickens can adversely impact the metabolism function [73]. Exposure to endosulfan in the eggs of a South American caiman has been reported to cause a reduction in egg weight and the hatchlings weight due to disruption of the metabolism of the embryo and the signals that control development [75]. Likewise, Endosulfan sulfate has been known to inhibit inhibited progesterone receptor binding in alligators [76].

Table 1. Physical and chemical properties of endosulfan and its metabolite [86]

Properties	α - Endosulfan	β - Endosulfan	Endosulfan Sulphate	Endosulfan diol
Molecular Weight (MW)	406.95 g	406.95 g	422.924 g	406.90 g
Molecular Formula	$C_9H_9C_{l6}O_3S$	$C_9H_9C_{l6}O_3S$	$C_9H_9C_{l6}O_4S$	$C_9H_6C_{l6}O_3S$
Vapor pressure (Vp) Pa at 25°C	6.0×10^{-3}	3.0×10^{-3}	1.0	-
Henry's law constant (H) Pam³mol⁻¹	10.23	1.94	2.61	-
Water solubility(Sw) mg/L at 25°C	0.32	0.33	0.117	0.33
Partition coefficient (log Kow)	3.83	3.62	3.66	-
Sorption coefficient (log Koc)	3.46	3.83	-	-
Melting point (Mp) (°C)	108-110	210-215	181	70-100
Density (g/cm³) at 25°C	-	-	1.94	1.745
State	Solid	Solid	Solid	Solid

4.2.4. Impacts on Soil and Sediments

Endosulfan is persistent in the soil media. It can be oxidized in soil by microbial activity to form endosulfan sulfate and by hydrolysis to form endosulfan-diol. Soil residues are the essential part of aquatic systems that represents a major long-term capacitor for holding pollutants. Reversibility of the pollutants interchanges between sediment and the covering water column results. In soil residues, the dynamics of accumulation of POPs is very complex. The accumulation of OCP residues in surface sediments of river, lake, and ocean may reflect fresh pollutants, though OCP residues in a very deep layer of stable soil residues cores provide past record of aquatic system exposure [77].

5. Residual Levels of Endosulfan in Indian Environment

5.1. Air

India is one of the largest consumers of endosulfan in the agriculture sector. The contamination of endosulfan has been reported in different part of India. The range of endosulfan measured in air of different Indian cities has been discussed in Table 2. Some megacities of India, namely Kolkata, Mumbai, Chennai, and Pondicherry reported a high level of α-endosulfan in the air [78]. In contrast, the remotely located city is comparatively less contaminated. Patna is one of the densely populated city in India whereas high as 2530 pg/m^3 of endosulfan in the air has been reported due to extensive use in the agriculture sector.

5.2. Water

Endosulfan concentration has been measured in surface water and groundwater by various researchers. It is opined that groundwater is less contaminated with endosulfan than surface water. High level of α-endosulfan (78ng/L) was measured in groundwater from Yavatmal region [79]. Likewise, surface water from Amrawati contained 42 ng/L of endosulfan concentration up to [80]. The residual level of endosulfan measured in different types of water bodies in the different part of India has been discussed in Table 3. In Gurgaon, the concentration of endosulfan was detected as high as 266.4ng/L in groundwater [81]. The open well in Thiruvallur measured 4518 ng/L of endosulfan [82]. Similarly, groundwater samples from Hyderabad showed a concentration of endosulfan ranging from 5917-8879 ng/L [83].

Table 2. Endosulfan concentration in air (pg/m^3)

Indian Cities	α-Endosulfan	References
Chennai	680	Zhang et al. [87]
Mumbai	498	
Baroda	317	
Kolkata	372	
Bangalore	3	
Cuddalore	185	
Portonovo	992	
Pichavaram	4	
Gulf of Kutch	12	
Thane Creek	1	Pandit and Sahu [88]
Parangipettai	-	Rajendran and Subramanian [89]
Tamil Nadu	−	Srimura et al. [90]
Imphal	30	Devi et al. [91]
Thoubal	89	
Waithou	80	
Laksar	972	Pozo et al. [92]
Sukanda	184	
Patna	2530	
Mudhol	630	
New Delhi	230	Chakraborty et al. [93]
Agra	340	
Goa	390	

Table 3. Endosulfan concentration in water sample (ng/L)

Indian Cities	Nature of water bodies	α-Endosulfan	Total Endosulfan	References
Kasargod, Kerala	Open Wells	-	<1000	Atmakuru and Ambalatharasu [94]
Bhandara	Groundwater	Nd–720	-	Lari et al. [95]
Bhandara	Surface water	Nd–80	-	
Amravati	Groundwater	Nd–60	-	
Amravati	Surface water	Nd–42	-	
Yavatmal	Groundwater	Nd–78	-	
Ambala	Groundwater	13.1	-	Kaushik et al. [96]
Gurgaon	Groundwater	266.4	-	
Hisar	Tap water	51.7	-	
Pilbhit	Gomti River	9.64	-	Malik et al. [97]
Chennai	Ennore Creek	29.21	-	Sunder [98]
Haryana	Groundwater	180	-	Kumara et al. [99]
Thiruvallur	Open Wells	4518	-	Jayshree and Vasudevan [100]
Unao	Stream, Ponds	Nd–130	-	Singh et al. [101]
Lucknow	Rain Water	2.99	-	Malik et al. [102]
Hyderabad	Groundwater	5917–8879	-	Shukla et al. [103]

5.3. Soil and Sediment

Soil and sediments are the important sinks for various types of organic pollutants including endosulfan. The level of endosulfan ranging from 1.78ng/g to 178ng/g in soil has been reported in different parts of India [84]. The concentration of endosulfan in soil media are represented in (Table 4). The residual level of endosulfan in soil from Haryana varied between 0.002 to 0.039 mg/g [85]. A range of 89,600–238,000 ng/g was measured in the soil residues of prawn ponds in Andhra Pradesh [69]. The Hooghly estuary in Eastern India detected 400mg/kg of endosulfan in sediment [86].

Higher concentration of α-endosulfan in Keoladeo has been also documented. However, the Ramnagara, in Karnataka had a low level of endosulfan. The Total endosulfan concentration is higher in Kasargod, Kerala but the lower concentration in Hasan, Karnataka.

Table 4. Endosulfan concentration in soil (mg/kg)

Indian Cities	α-Endosulfan	β-Endosulfan	Endosulfan Sulphate	Total Endosulfan	References
Ramanagara, Karnataka	0.2105	0.515	8.52	9.24	Sunitha et al. [104]
Hassan, Karnataka	-	0.075	1.068	1.143	
Hommagarahalli, Mysore	-	-	2.96	2.96	
Kasargod, Kerala	3.731	21.424	25.094	50.25	
Hyderabad	18050	-	-	-	Kata et al. [105]
Assam	19-41	-	-	-	NRCC [106]
Tripura	9-57	-	-	-	NRCC [106]
Manipur	20-470	-	-	-	NRCC [106]
Andaman Island	4790	-	-	-	Murugan et al. [107]
Keoladeo	4569000	-	-	-	Bhadouria et al. 108]
NCR Delhi	950	-	-	-	Kumar et al. [109]
Gomti River	130	-	-	-	Singh [110]
Hissar	2000–39000	-	-	-	Kumari et al. [99]
Unnao	Nd–13070	-	-	-	Singh et al. [101]
Agra	18050	-	-	-	Singh [110]
West Bengal	-	-	400	-	Bhattacharya et al. [85]

5.4. Biota

The concentration of endosulfan in Kerala state of different biotic species like fish, milk, and leaves has been detected in greater than 1000µg/kg [94].

SUMMARY AND CONCLUSION

Endosulfan is one of the highly toxic organochlorine insecticides and also one of the dirty dozen compound listed by Stockholm convention of POPs. Endosulfan is lethal to all types of organism and the environment. India is one of the leading consumer and producer of endosulfan in the agriculture sector. A number of studies have detailed endosulfan contamination in different environmental media, for instance, air, water, soil/sediments, and biota. This study highlighted the high residual level of endosulfan in the air, water, soil/sediments and biota of India despite its low intake rate in the recent year.

REFERENCES

[1] Halsall, C. J. 2004. Investigating the occurrence of persistent organic pollutants (POPs) in the Arctic: their atmospheric behaviour and interaction with the seasonal snow pack. *Environ Pollut.* 128.

[2] Halsall, C. J., Bailey, R., Stern, G., Barrie, L. A., Fellin, P., Muir, D. C G., et al. 1998. Multi-year observations of organohalogen pesticides in the Arctic atmosphere. *Environ Pollut.* 102.

[3] Maier-Bode, H. 1968. Properties, effect, residues and analytics of the insecticide endosulfan. *Residue Rev.* 22.

[4] Herrmann, M. 2002. *Preliminary risk profile of endosulfan.* Berlin Germany: Umweltbundesamt.

[5] Roberts, D. M., Karunarathna, A., Buckley, N. A., Manuweera, G., Sheriff, M. H. R., Eddleston, M. 2003. Influence of pesticide regulation on acute poisoning deaths in Sri Lanka. *Bull World Health Organ.* 8.

[6] Gregor, D. J. 1990. Deposition and accumulation of selected agricultural pesticides in Canadian Arctic snow. In: Kurtz DA, editor. *Long range transport of pesticides.* Chelsea, Mich.: Lewis Publishers.

[7] Bidleman, T. F., Patton, W. W., Hinckley, D. A., Walla, M. D., Cotham, W. E., Hargrave, B T. 1990. Chlorinated pesticides and polychlorinated biphenyls in the atmosphere of the Canadian Arctic. In: Kurtz DA, editor. *Long Range Transport of Pesticides.* Chelsea, Mich.: Lewis Publishers.

[8] Bidleman, T. F., Cotham, W. E., Addison, R. F., Zinck, M. E. 1992. Organic contaminants in the northwest Atlantic atmosphere at Sable Island, Nova Scotia, 1988–1989. *Chemosphere.* 24.

[9] Pozo, K., Harner, T., Wania, F., Muir, D. C. G., Jones, K. C., Barrie, L A. 2006. Toward a global network for persistent organic pollutants in air: results from the GAPS study. *Environ Sci Technol.* 40.

[10] Usenko, S, Landers, D H., Appleby, P G., Simonich, S L. 2007. Current and historical deposition of PBDEs, pesticides, PCBs, and PAHs to Rocky Mountain national park. *Environ Sci Technol.* 41.

[11] Rand, G. M., John, F., Carriger, Piero R., Gardinali, and Joffre Castro. 2010. "Endosulfan and its metabolite, endosulfan sulfate, in freshwater ecosystems of South Florida: a probabilistic aquatic ecological risk assessment." *Ecotoxicology* 19, no. 5.

[12] Stanley, K. A., Lawrence, R., Curtis, Staci L., Massey, S., and Robert, L. T. 2009. "Endosulfan I and endosulfan sulfate disrupts zebrafish embryonic development." *Aquatic toxicology* 95, no. 4. (Stanley et al. 2009, 355-361).

[13] Ramesh, A., Tanabe, S., Tatsukawa, R., Subramanian, A. N., Palanichamy, S., Mohan, D., & Venugopalan, V. K. 1989. Seasonal variations of organochlorine insecticide residues in air from Porto Novo, South India. *Environmental Pollution*, 62(2-3). (Ramesh et al. 1989, 213-222).

[14] Ramesh, A., Tanabe, S., Subramanian, A. N., Mohan, D., Venugopalan, V. K., & Tatsukawa, R. 1990. Persistent organochlorine residues in green mussels from coastal waters of South India. *Marine pollution bulletin*, 21(12). (Ramesh 1990, 587-590).

[15] Iwata, H., Shinsuke, T., and Tatsukawa, R. 1994. "Global contamination by persistent organochlorines and their ecotoxicological impact on marine mammals." *Science of the total environment 154*, no. 2-3. (Iwata et al. 1994, 163-177).

[16] Iwata, H., Tanabe, S., Sakai, N., and Tatsukawa, R. 1993. "Distribution of persistent organochlorines in the oceanic air and surface seawater and the role of ocean on their global transport and fate." *Environmental Science & Technology* 27, no. 6. (Iwata et al. 1993, 1080-1098).

[17] Senthilkumar, K., Duda, C. A., Villeneuve, D. L., Kannan, K., Falandysz, J., & Giesy, J. P. 1999. Butyltin compounds in sediment and fish from the Polish coast of the Baltic Sea. *Environmental Science and Pollution Research*, 6(4), (Senthilkumar et al. 1999, 200).

[18] Tanabe, S., Subramanian, A. N., Ramesh, A., Kumaran, P. L., Miyazaki, N., & Tatsukawa, R. 1993. Persistent organochlorine residues in dolphins from the Bay of Bengal, South India. *Marine Pollution Bulletin*, 26(6). (Tanabe et al. 1993, 311-316).

[19] Rajendran, R. Babu, and Subramanian, A. N. 1999. "Chlorinated pesticide residues in surface sediments from the River Kaveri, South India." *Journal of Environmental Science & Health Part B* 34, no. 2. (Rajendran and Subramanian 1999, 269-288).

[20] Zhang, G., Chakraborty, P., Li, J., Sampathkumar, P., Balasubramanian, T., Kathiresan, K., Takahashi, S., Subramanian, A., Tanabe, S., and Jones, Kevin C. 2008. "Passive atmospheric sampling of organochlorine pesticides, polychlorinated biphenyls, and polybrominated diphenyl ethers in urban, rural, and wetland sites along the coastal length of India." *Environmental science & technology* 42, no. 22. (Zhang et al.. 2008, 8218-8223).

[21] Ramesh, A., Tanabe, S., Murase, H., Subramanian, A. N., and Tatsukawa, R.1991. "Distribution and behaviour of persistent

organochlorine insecticides in paddy soil and sediments in the tropical environment: a case study in South India." *Environmental Pollution* 74, no. 4.

[22] Tanabe, S., Prudente, M., Mizuno, T., Hasegawa, J., Iwata, H., and Miyazaki, N. 1998. "Butyltin contamination in marine mammals from North Pacific and Asian coastal waters." *Environmental science & technology32*, no. 2.

[23] Senthilkumar, kumar., Kannan, K., Subramanian, A., and Tanabe, S. 2001. "Accumulation of organochlorine pesticides and polychlorinated biphenyls in sediments, aquatic organisms, birds, bird eggs and bat collected from South India." *Environmental Science and Pollution Research* 8, no. 1.

[24] Gregor, D. J., and Gummer, W. D. 1989. Evidence of atmospheric transport and deposition of organochlorine pesticides and polychlorinated biphenyls in Canadian Arctic snow. *Environ Sci Technol.* 23. (Gregor and Gummer 1989, 561–565).

[25] Bidleman, T. F., Cotham, W. E., Addison, R. F., Zinck, M. E. 1992. Organic contaminants in the northwest Atlantic atmosphere at Sable Island, Nova Scotia, 1988–1989. *Chemosphere.* 24. (Bidleman et al. 1992, 1389–412).

[26] Hoff, R. M., Muir, D. C. G., Grift, N. P. 1992. Annual cycle of polychlorinated biphenyls and organohalogen pesticides in air in Southern Ontario. 2. Atmospheric transport and sources. *Environ Sci Technol* .26. (Hoff et al. 1992, 276–83).

[27] Shen, L., Wania, F., Ying, D. L., Teixeira, C., Muir, D. C. G., Bidleman, T F. 2005. Atmospheric distribution and long-range transport behaviour of organochlorine pesticides in North America. *Environ Sci Technol* .39. (Shen et al. 2005, 409–20).

[28] DPR. 2004. Summary of Physical and Chemical Properties. Pesticides Database. Environmental Monitoring Branch, Department of Pesticide Regulation, CalEPA. Sacramento, California. http://dpr01.inside.cdpr.ca.gov:8000/cgi-bin/cheminfo/cheminfo.pl (DPR 2004).

[29] UNEP. Final act of the plenipotentiaries on the Stockholm Convention on persistent organic pollutants. United Nations environment program chemicals. Switzerland: Geneva. (UNEP 2001, 445).

[30] US EPA. 2002. Endosulfan RED Facts. US EPA, Office of Prevention, Pesticides. http://www.epa.gov/oppsrrd1/REDs/factsheets/endosulfan_fs.htm (US EPA 2002).

[31] Wan, M. T., Szeto, S., Price, P. 1995. Distribution of Endosulfan residues in the drainage waterways of the lower Fraser valley of British-Columbia. *J Environ Sci Health B*. 30. (Wan et al. 1995, 401–33).

[32] Kathpal, T. S., Singh, A., Dhankhar, J. S., Singh, G. 1997. Fate of endosulfan in cotton soil under subtropical conditions of northern India. *Pestic Sci*. 50.

[33] GFEA (German Federal Environment Agency). 2004. Draft Dossier prepared in support of a proposal of endosulfan to be considered as a candidate for inclusion in the UN-ECE LRTAP protocol on persistent organic pollutants. German Federal Environment Agency. Umweltbundesamt, Berlin. http://www.unece.org/env/popsxg/docs/2004/Dossier_Endosulfan.2004.pdf (GFEA 2004).

[34] Bedose C., Cellier P., Calvet R., Barriuso E. 2002. Occurrence of pesticides in the atmosphere in France. *Agronomie* 22. (Bedose 2002, 35-49)

[35] ATSDR. 2000. Toxicological Profile for Endosulfan. Agency of Toxic Substances and Disease Registry, Atlanta, USA. http://www.atsdr.cdc.gov/toxprofiles/tp41.html (ATSDR. 2000).

[36] Hageman K. J, Simonich S. L., Campbell D. H., Wilson G. R, Landers DH. 2006. Atmospheric deposition of current-use and historic-use pesticides in snow at national parks in the western United States. *Environ Sci Technol* 40. (Hageman 2006, 3174-80).

[37] Daly G., Lei Y. D., Teixeira C., Muir D. C. G., Castillo L. E., Wania F. 2007b. Accumulation of current-use pesticides in neotropical montane forests. *Environ Sci Total* 41. (Daly 2007b, 118-23).

[38] Tuduri L., Harner T., Blanchard P., Li Y. F., Poissant L., Waite D. T., Murphy C, Belzer W. 2006. A review of currently used pesticides

(CUPs) in Canadian air and precipitation: Part 1: Lindane and endosulfans. *Atmos Environ* 40:1563-78. (Tuduri 2006, 1563-78).
[39] GFEA-U. 2007. Endosulfan. Draft Dossier prepared in support of a proposal of endosulfan to be considered as a candidate for inclusion in the CLRTAP protocol on persistent organic pollutants. German Federal Environment Agency – Umweltbundesamt, Berlin. (GFEA-U. 2007).
[40] Bollmohr S., Day J. A., Schulz R. 2007. Temporal variability in particle associated pesticide exposure in a temporarily open estuary, Western Cape, South Africa. *Chemosphere* 68. (Bollmohr 2007, 479-88).
[41] Schmidt, W. F., Bilboulian, S., Rice, C. P., Fettinger, J. C., McConnell, L. L., Hapeman, C. J. 2001. Thermodynamic, spectroscopic, and computational evidence for the irreversible conversion of alpha to beta-endosulfan. *J Agric Food Chem*.49. (Schmidt et al. 2001, 5372–6).
[42] Walse, S. S., Shimizu, K. D., Ferry, J. L. 2002. Surface-catalyzed transformations of aqueous endosulfan. *Environ Sci Technol*. 36. (Walse et al. 2002, 4846–53).
[43] Rice, C. P., Chernyak, S. M., McConnell, L. L. 1997. Henry's Law constants for pesticides measured as a function of temperature and salinity. *J Agric Food Chem*. 45. (Rice et al. 1997b, 2291–5).
[44] Moon J. M., Chun B. J. 2009. Acute endosulfan poisoning: a retrospective study. Hum Exp Toxicol. 28. (Moon 2009, 309-16).
[45] Satar S., Sebe A., Alpay N. R., Gumusay U., Guneysel O. 2009. Unintentional endosulfan poisoning. *Bratisl LekListy*.;110(5). (Satar 2009, 301-5).
[46] Anonymous., 2008 Final report of the investigation of unusual illnesses allegedly produced by endosulfan exposure in padre village of Kasargod district (N.Kerala). National Institute of occupational Health (Indian Council of Medical Research). (Anonymous 2002, 98).
[47] Silva M. H., Gammon D., 2009. An assessment of the developmental, reproductive and neurotoxicity of endosulfan. *Birth Defects Research*

Part B: *Developmental and Reproductive Toxicology*, 86. (Silva 2009, 1-28).

[48] Silva M. H. Endosulfan risk characterization document. Medical toxicology and worker health and safety branches department of pesticide regulation California environmental protection. 2007; Available from: www.cdpr.ca.gov/docs/emon/pubs/tac/.../endosulfan/ endosulfan_doc.pdf. (Silva 2007).

[49] Kishi M. 2002 Acutely Toxic pesticides. Report submitted to IFCS Workgroup. International Forum On Chemcial Safety. http://www.who.int/heli/risks/toxics/bibliographyikishi.pdf.

[50] NIOH. 2003. Final Report of the Investigation of Unusual Illnesses Allegedly Produced by Endosulfan Exposure In Padre Village of Kasargod District (N Kerala). National Institute of Occupational Health, Indian Council for Medical Research, Ahmedabad. (NIOH 2003).

[51] Wesseling C., Corriols M., Bravo V. 2005. Acute pesticide poisoning and pesticide registration in Central America. *Toxicol Appl* Pharmacol 207(2 Suppl 1):697-705. (Wesseling 2005, 697-705).

[52] Glin L. J., Kuiseau J., Thiam A., Vodouhe D. S., Dinham B., Ferrigno S. 2006. Living with Poison: Problems of Endosulfan in West Africa Cotton Growing Systems. Pesticide Action Network UK, London.(Glin 2006)

[53] Karatas A. D., Aygun D., Baydin A., 2006. Characteristics of endosulfan poisoning: a study of 23 cases. *Singapore Medical Journal*, 47. (Karatas 2006, 1030-1032).

[54] Tietz W., 1999. *Fundamentals of Clinical Chemistry* (Eds. CA. Burtis and ER. Ashwood) W. B. Saunders Comp. Philadelphia, London, Toronto, Montreal, Sydney, Tokyo. (Tietz 1999, 803-804).

[55] Anonymous., 2000. Toxicological profile for endosulfan. U. S. Department of health and human services. Public health service. Agency for toxic substances and disease registry. (Anonymous 2000, 323).

[56] Sunol C., Babot Z., Fonfria E., Galofre M., Garcia D., Herrera N., Iraola S., Vendrell I., 2008. Studies with neuronal cells: from basic

studies of mechanisms of neurotoxicity to the prediction of chemical toxicity. *Toxicol In Vitro*, 22. (Sunol 2008, 1350–1355).

[57] Wozniak A. L., Bulayeva N. N., Watson C. S. Xenoestrogens at picomolar to nanomolar concentrations trigger membrane estrogen receptor-alpha-mediated Ca^{2+} fluxes and prolactin release in GH3/B6 pituitary tumor cells. *Environmental Health Perspectives*, 113. (Wozniak 2005, 431–439).

[58] Chen L., Durkin K. A., Casida J. E. Structural model for gamma-aminobutyric acid receptor noncompetitive antagonist binding: widely diverse structures fit the same site. *Proceedings of the National Academy of Sciences of the United States of America*, 103. (Chen 2006, 5185–5190).

[59] Cole L. M., Casida J. E. Polychlorocycloalkane insecticide-induced convulsions in mice in relation to disruption of the GABA-regulated chloride ionophore. *Life Sciences*, 39. (Cole 1986, 1855–1862).

[60] Bajpayee M., Pandey A., Zaidi S., Musarrat J., Parmar D., Mathur N., Seth P., Dhawan A. DNA damage and mutagenicity induced by endosulfanand and its metabolites. *Environmental and Molecular Mutagenesis*, 47. (Bajpayee 2006, 682-692).

[61] Spencer, P. S., and Schaumburg, H. H. 2000. Chlorinated cyclodienes. In: Spencer, P. S., Scheumburg, H. H., Ludolph, A. C., eds.; Oxford University Press: Oxford. (Spencer and Schaumburg 2000).

[62] Weber, J., Halsall, C J., Muir D., Camilla, T., Jeff, S., Keith, S., Mark, H., Hayley, H., Terry, B. 2009. "Endosulfan, a global pesticide: A review of its fate in the environment and occurrence in the Arctic". *Science of the Total Environment*. 408 (15): doi:10.1016/j. scitotenv. 0.077. PMID 19939436, 20107. (Weber et al. 2009, 2966-2984).

[63] Dutta H., Arends D. A. 2003. Effects of endosulfan on brain acetylcholinesterase activity in juvenile bluegill sunfish. *Environ Res* 91. (Dutta 2003, 157-62).

[64] Naqvi S. M., Vaishnavi C. 1993. Bioaccumulative potential and toxicity of endosulfan insecticide to non-target animals. *Comp Biochem Physiol C* 105(3). (Naqvi 1993, 347-61).

[65] Sharma S., Nagpure N. S., Kumar R., Pandey S., Srivastava S. K., Singh PJ, Mathur PK. 2007a. Studies on the genotoxicity of endosulfan in different tissues of fresh water fish *Mystus vittatus* using the Comet assay. *Arch Environ Contam Toxicol* 53(4):617-23. (Sharma 2007a, 617-23).

[66] Wessel N., Rousseau S., Caisey X., Quiniou F., Akcha F. 2007. Investigating the relationship between embryotoxic and genotoxic effects of benzo[a]pyrene, 17 *a*-ethinylestradiol and endosulfan on *Crassostrea gigas* embryos. *Aquat Toxicol* 85. (Wessel 2007, 133-42).

[67] Ffrench-Constant, R. H., Anthony, N., Aronstein, K., Rocheleau, T., Stilwell, G. 2002. Cyclodiene insecticide resistance: from molecular to population genetics. *Annual Reviews Entomology*.45. (Ffrench-Constant et al. 2002, 449–466).

[68] Bretaud, S., Toutant, J. P., Saglio, P. 2002. Effects of carbofuran, diuron, and nicosulfuron on acetylcholinesterase activity in goldfish (Carassius auratus). *Ecotoxicology and Environmental Safety*. 47. (Bretaud et al. 2002, 117–124).

[69] Estellano, V. R. H., Pozo, K., Harner, T., Franken, M., Zaballa, M. 2008. Altitudinal and seasonal variations of persistent organic pollutants in the Bolivian Andes mountains. *Environ. Sci. Technol*. 42. (Estellano 2008, 2528–2534).

[70] https://www.google.com/search?q=process+of+endosulfan+move+into+soil+crops+image&tbm=isch&tbs=rimg:CadVawaANx_1IjguVOMJGTlEH2vzI0c57sbQrBH8Mu4A4y4pn6RaWexm1VtU7kgy_1BzI8dn1gImUiyQ5Xo_1RKh1ZioSCS5U4wkZOUQfEQNquRgiI4PaKhIJa_1MjRznuxtAR_1kMYzgBwpS8qEgmsEfwy7j4DjBEC9jQeixk8UCoSCbimfpFpZ7GbEZMchBE_1kD93KhIJVW1TuSDL8HMRT1ezP0jtGGoqEgkjx2fWAiZSLBHMayxBva4yJCoSCZDlej9EqHVmEaN4bFoDGhM&tbo=u&sa=X&ved=2ahUKEwjPlPvR35nhAhUTY48KHUURBNMQ9C96BAgBEBs&biw=1366&bih=625&dpr=1#imgrc=NPWQUg-lywJhAM:

[71] Naqvi S. M., Vaishnavi C. 1993. Bioaccumulative potential and toxicity of endosulfan insecticide to non-target animals. *Comp Biochem Physiol C* 105(3).(Naqvi 1993, 347-61)

[72] GFEA-U. 2007. Endosulfan. Draft Dossier prepared in support of a proposal of endosulfan to be considered as a candidate for inclusion in the CLRTAP protocol on persistent organic pollutants. German Federal Environment Agency – Umweltbundesamt, Berlin. (GFEA-U. 2007)

[73] Garg U. K., Pal A. K., Jha G. J., Jadhao S. B. 2004. Haemato-biochemical and immuno-pathophysiological effects of chronic toxicity with synthetic pyrethroid, organophosphate and chlorinated pesticides in broiler chicks. *Int Immunopharmacol* 4(13). (Garg 2004, 1709-22)

[74] Pushpanjali, Pal A. K., Prasad R. L., Prasad A., Singh S. K., Kumar A, Jadhao S. B. 2005. In ovo embryotoxicity of a-endosulfan adversely influences liver and brain metabolism and the immune system in chickens. *Pestic Biochem Physiol* 82. (Pushpanjali 2005, 103–14)

[75] Beldomenico P. M., Rey F., Prado W. S., Villarreal J. C., Muñoz-de-Toro M, Luque EH. 2007. In ovum exposure to pesticides increases the egg weight loss and decreases hatchlings weight of Caiman latirostris (Crocodylia: Alligatoridae). *Ecotox Environ Saf* 68(2). (Beldomenico 2007, 246-51)

[76] Vonier P. M., Crain D. A., McLachlan J. A., Guilette L. J. Jr, Arnold SF. 1996. Interaction of environmental chemicals with the estrogen and progesterone receptors from the oviduct of the American alligator. *Enviro Health Perspect* 104(12).(Vonier 1996, 1318-22)

[77] Amaraneni, S. R. 2006. Distribution of pesticides, PAHs and heavy metals in prawn ponds near Kolleru lake wetland, India. Environ Int, 32. (Amaraneni 2006, 294 -302).

[78] Chakraborty P., Zhang G., Li J., Xu Y., Liu X., Tanabe S. et al. 2010. Selected organochlorine pesticides in the atmosphere of major Indian cities: levels, regional versus local variations, and sources. *Environmental Science and Technology*, 44. (Chakraborty et al. 2010, 8038–8043).

[79] Lari, S. Z., Khan, N. A., Gandhi, K. N., Meshram, T. S., Thacker, N. P., 2014. Comparison of pesticide residues in surface water and ground water of agriculture intensive areas. *J. Environ. Health Sci. Eng.* 12. (Lari 2014, 11–19).

[80] Kaushik, C. P., Sharma, H. R., Kaushik, A., 2012. Organochlorine pesticide residues in drinking water in the rural areas of Haryana, India. Environ. *Monit. Assess.* 184. (Kaushik 2012, 103–112).

[81] Jayashree, R., Vasudevan, N., 2005. Residues of organochlorine pesticides in agricultural soils of Thiruvallur district, India. *Food. Agric. Environ.* 4 (1).(Jayashree 2005, 313–316).

[82] Shukla, G., Kumar, A., Bhanti, M., Joseph, P. E., Taneja, A., 2006. Organochlorine pesticide contamination of ground water in the city of Hyderabad. *Environ. Int.* 32. (Shukla 2006, 244–247).

[83] Shukla, Gangesh, Anoop Kumar, Mayank Bhanti, P. E. Joseph, and Ajay Taneja. 2006. "Organochlorine pesticide contamination of ground water in the city of Hyderabad." *Environment international* 32, no. 2. (Shukla 2006, 244-247).

[84] Kumari, Beena, V. K. Madan, and T. S. Kathpal. 2008. "Status of insecticide contamination of soil and water in Haryana, India." *Environmental monitoring and assessment* 136, no. 1-3. (Kumari 2008, 239-244).

[85] Bhattacharya S., Gosh R. K., Mandal T. K., Chakraborty A. K., Basak DK. 1993. Some histological changes in chronic endosulfan (Thionol) toxicity in poultry. *Indian J Anim Health* 32. (Bhattacharya 1993, 9-11).

[86] Dorough, H. Wyman, Kurt Huhtanen, Thomas C. Marshall, and Harry E. Bryant. 1978. "Fate of endosulfan in rats and toxicological considerations of apolar metabolites." *Pesticide Biochemistry and Physiology* 8, no. 3. (Dorough 1978, 241-252).

[87] Zhang, G., Chakraborty, P., Li, J., Sampathkumar, P., Balasubramanian, T., Kathiresan, K., Takahashi, S., Subramanian, A., Tanabe, S., and Jones, K. C., 2008. Passive atmospheric sampling of organochlorine pesticides, polychlorinated biphenyls, and polybrominated diphenyl ethers in urban, rural, and wetland sites

along the coastal length of India, *Environ. Sci. Technol.*, 42. https://doi.org/10.1021/es8016667. (Zhang 2008, 8218–8223).

[88] Pandit, G. G., Sahu, S. K., 2001. Gas exchange of OCPs across the air–water interface at the creek adjoining Mumbai harbor, *India. J. Environ. Monit.* 3. (Pandit 2001, 635–638).

[89] Rajendran, R. B., Subramanian, A. N., 1997. Pesticide residues in water from Kaveri, South India. *Chem. Ecol.* 13. (Rajendran 1997, 57–70).

[90] Srimural, S., Govindaraj, S., Kumar, S. K., Rajendran, R. B., 2014. Distribution of organochlorine pesticides in atmospheric air of Tamil Nadu, Southern India. *Int. J. Environ. Sci. Technol.* http://dx.doi.org/10.1007/s13762-014-0558-3 (in press). (Srimural 2014).

[91] Devi, N. L., Qi, S., Chakraborty, P., Zhang, G., Yadav, I. C., 2011. Passive air sampling of organochlorine pesticides in a northeastern state of India, Manipur. *J. Environ. Sci.* 23 (5).(Devi 2011, 808-815).

[92] Pozo, K., Harner, T., Lee, S. C., Sinha, R. K., Sengupta, B., Loewen, M., Geethalakshmi, V., Kannan, K., Volpi, V., 2011. Assessing seasonal and spatial trends of persistent organic pollutants (POPs) in Indian agricultural regions using PUF disk passive air samplers. *Environ. Pollut.* 159. (Pozo 2011, 646-653).

[93] Chakraborty, P., Zhang, G., Li, J., Xu, Y., Liu, X., Tanabe, S., Jones, K. C., 2010. Selected organochlorine pesticides in the atmosphere of major Indian cities: levels, regional versus local variations, and sources. *Environ. Sci. Technol.* 44. (Chakraborty 2010, 8038–8043).

[94] Atmakuru, Ramesh and Ambalatharasu, Vijayalakshmi, 2002. "Environmental exposure to residues after aerial spraying of endosulfan: residues in cow milk, fish, water, soil and cashew leaf in Kasargode, Kerala, India Pest Management Science Pest Manag Sci 58. DOI: 10.1002/ps.568.(Atmakuru 2002,1048-1054)

[95] Lari, S. Z., Khan, N. A., Gandhi, K. N., Meshram, T. S., Thacker, N.P., 2014. Comparison of pesticide residues in surface water and ground water of agriculture intensive areas. *J. Environ. Health Sci. Eng.* 12.(Lari 2014, 11–19).

[96] Kaushik, C. P., Sharma, H. R., Kaushik, A., 2012. Organochlorine pesticide residues in drinking water in the rural areas of Haryana, India. *Environ. Monit. Assess.* 184. (Kaushik 2012, 103–112).
[97] Malik, A., Ojha, P., Singh, K. P., 2009. Levels and distribution of persistent organochlorine pesticide residues in water and sediments of Gomti River (India)—a tributary of the Ganges River. *Environ. Monit. Assess.* 148. (Malik 2009, 421–435).
[98] Sundar, G., Selvarani, J., Gopalakrishnan, S., Ramachandran, S., 2010. Occurrence of organochlorine pesticide residues in green mussel (Perna viridisL.) and water from Ennore creek, Chennai, India. Environ. Monit. Assess. 160. (Sundar 2010, 593–604).
[99] Kumari, B., Madan, V. K., Kathpal, T. S.,2008. Status of insecticide contamination of soil and water in Haryana, India. *Environ. Monit. Assess.* 136. (Kumari 2008, 239–244).
[100] Jayashree, R., Vasudevan, N., 2005. Residues of organochlorine pesticides in agricultural soils of Thiruvallur district, India. *Food. Agric. Environ.* 4 (1). (Jayashree 2005, 313–316).
[101] Singh, K. P., Malik, A., Sinha, S., 2007. Persistent organochlorine pesticide residues in soil and surface water of northern Indo-Gangetic alluvial plains. *Environ Monit Assess* 125. (Singh 2007, 147–155).
[102] Malik, A., Singh, V. K., Singh, K. P., 2007. Occurrence and Distribution of Persistent Trace Organics in Rainwater in an Urban Region (India). Bull. Environ. *Contam. Toxicol.* 79. (Malik 2007, 639–645).
[103] Shukla, G., Kumar, A., Bhanti, M., Joseph, P.E., Taneja, A., 2006. Organochlorine pesticide contamination of ground water in the city of Hyderabad. *Environ. Int.* 32. (Shukla 2006, 244–247).
[104] Sunitha, Sarah, Krishnamurth, V. and Mahmood, Riaz. Analysis of Endosulfan residues in cultivated soils in Southern India. *International Conference on Biotechnology and Environment Management, vol. 18.*(Sunitha *2011, 110-114)*
[105] Kata, M., Rao, S. S., Mohan, K. R., 2014. Spatial distribution, ecological risk evaluation and potential sources of organochlorine

pesticides from soils in India. Environ. Earth Sci. http://dx.doi.org/10.1007/s12665-014-3189-6 (in press). (Kata 2014).

[106] National Research Council Canada. 1975. Associate Committee on Scientific Criteria for Environmental Quality. Subcommittee on Pesticides, and Related Compounds. *Endosulfan: its effects on environmental quality*. Vol. 14098. Nationl Research Council, Canada. (National Research Council Canada 1975).

[107] Murugan, A. V., T. P. Swarnam, and S. Gnanasambandan., 2013. "Status and effect of pesticide residues in soils under different land uses of Andaman Islands, India." *Environmental monitoring and assessment* 185, no. 10. (Murugan 2013, 8135-8145).

[108] Bhadouria, B. S., Mathur, B. V., Kaul, R., 2012. Monitoring of organochlorine pesticides in and around Keoladeo National Park, Bharatpur, Rajasthan, India. *Environ. Monit. Assess*. 184. (Bhadouria 2012, 5295–5300).

[109] Kumar, B., Kumar, S., Gaur, R., Goel, G., Mishra, M., Singh, S. K., Prakash, D., Sharma, C. S., 2011. Persistent Organochlorine Pesticides and Polychlorinated Biphenyls in Intensive Agricultural Soils from North India. *Soil & Water Res*. 6 (4). (Kumar 2011, 190–197.

[110] Singh, R. P., 2001. Comparison of Organochlorine Pesticide Levels in Soil and Groundwater of Agra. India. *Bull. Environ. Contam. Toxicol*. 67. (Singh 2001, 126–132).

Chapter 3

ENDOSULFAN: PRACTICES, POLICIES, AND ECO-TOXICOLOGICAL IMPACTS IN INDIA

*Sanjenbam Nirmala Khuman[1], Girija Bharat[2],
Avanti Roy-Basu[2], Piyush Mohapatro[3]
and Paromita Chakraborty[1],**

[1]Department of Civil Engineering, SRM Research Institute, SRM Institute of Science and Technology, Kattankulathur, Tamil Nadu, India
[2]Mu Gamma Consultants Pvt. Ltd., Gurgaon, India
[3]Toxics Link, Jungpura Extension, New Delhi, India

ABSTRACT

Endosulfan has been widely used as an insecticide in India. Endosulfan is one of the toxic and persistent compounds in the environment among several organochlorine pesticides. This article detailed the characteristics of Endosulfan, its impact on human health and environment, and the policies and practices pertaining to Endosulfan usage in India. Additionally, Endosulfan residues in air/soil/water in India were also summarized to understand the current status of Endosulfan in the

---

* Corresponding Author's E-mail: paromita.c@res.srmuniv.ac.in.

environment. Eco-toxicological risk assessment based on previous literature showed harmful impacts even on edible fish species. In many places, Endosulfan residue in underground water exceeded the permissible limit and is not fit for drinking purpose. Following the negative impacts, this chapter also highlighted some alternatives recommended by the Indian government which can be used instead of Endosulfan. In many parts of India, Endosulfan has been replaced by biological pesticides such as neem seed-kernel extracts and chili–garlic extracts to control bollworms and sucking insects. The scientific community can play an active role in setting an agenda for management of Endosulfan and the protection of public health at large.

1. INTRODUCTION

Endosulfan is a broad-spectrum pesticide under organochlorines groups, which is used to control a number of insects on food crops such as food grains, tea, fruits, vegetables and other non-food crops such as tobacco and cotton [1]. Another application of Endosulfan is as a wood preservative. Technical Endosulfan mixture contains two isomers α and β-Endosulfan in the proportion of 7:3. Endosulfan is persistent in nature and has a photolytic half-life of about 7 days in air. The hydrolytic half-lives of Endosulfan under anaerobic conditions range between 35 to 37 days [1]. Endosulfan enters the environment during its production and application via volatilization in the air, leaching in water and adsorption on soil particles. Endosulfan in the air may travel long distances before it lands on crops, soil, or water. Rainwater washes Endosulfan from soil into surface water via surface runoff. Endosulfan is hydrophobic in nature and it is attached mostly to soil particles floating in the water or to sediments. Endosulfan breaks down into Endosulfan sulfate, Endosulfan diol, and Endosulfan furan.

In terms of toxicity, various research studies have shown negative health effects of using Endosulfan, such as neurotoxicity, late sexual maturity, physical deformities (also of newborns) and others [2]. The World Health Organization (WHO) has categorized Endosulfan as moderately hazardous (i.e., Class II), and the European Union (EU) has identified it to be: harmful when in contact with skin, very toxic by inhalation, very toxic if swallowed, very toxic/dangerous for the environment; very toxic to aquatic organisms, and may cause long term effects in the aquatic environment. There are many cases of Endosulfan poisoning reported in different countries, which ultimately lead to various health hazards, severe disabilities or even death.

In India, it has been widely produced and used for growing food crops and cash crops such as cashew, cotton, tea, paddy, fruits, etc. It is used to control agricultural pests like tea mosquito bug, thrips, mites, beetles, caterpillars, borers, cutworms, bollworms, bugs, whiteflies, leafhoppers, and snails. India began to produce Endosulfan in 1996 and had become the leading producer by 2004, with more than 60 Endosulfan manufacturers and formulators [3]. The three leading Endosulfan manufacturing companies were Coromandel Fertilisers Ltd, Hindustan Insecticides Ltd., and Excel Crop Care; the latter being the largest with a reported capacity of 6,000 tonnes per annum [4]. It is reported that in the year 2007-08, the top three manufacturers have produced a total of 9500 Metric Tonnes (MT) and exported 4000 MT of Endosulfan. With such huge production and usage history, it is important to understand the current scenario of Endosulfan in India and steps taken to control its usage. This chapter is therefore based on the following objectives:

a. Understanding the usage/practices of Endosulfan in India
b. Occurrences and eco-toxicological risk assessment of Endosulfan in India
c. Understanding the effects of Endosulfan through a case study
d. Alternatives/policies of Endosulfan in India
e. The road ahead: Key action points and recommendations

1.1. Usage Rules and Policy Compliances

In 2011, the Stockholm Convention (SC) on Persistent Organic Pollutants (POPs) announced a global ban on the production and use of Endosulfan, with certain uses exempted for five additional years. Prior to the worldwide ban on Endosulfan by SC on POPs, more than 80 countries had already announced ban or phase out of Endosulfan. In 2010, the United States banned the usage of Endosulfan. However, India did not agree to the ban on Endosulfan in Fifth Conference of Parties held in May 2011 by the SC. India opposed this ban, and asked for a remission of 10 years as the company owned by the Government of India was the major manufacturer of Endosulfan. The argument raised (by the Endosulfan manufacturing companies) was that it is one of the cheapest pesticides available in the world market as compared to its eco-friendly but expensive alternatives. The Pesticide Manufacturers and Formulators Association of India (PMFAI) even went ahead and termed it as a "conspiracy" by the European Union (EU) to promote the use of the costlier alternatives (EU promoted) of Endosulfan which will be unaffordable by poor Indian farmers.

Table 1. Timeline of Endosulfan management in India

Early 1950s	Endosulfan was developed globally
Early 1980s	Endosulfan was used in agriculture in many parts of India including aerial spraying in Kerala and Karnataka
1996	India began commercial production of Endosulfan
2001	The state of Kerala banned the use of Endosulfan
2004	India became the largest producer and user of Endosulfan
2010	The United States banned the use of Endosulfan
2011	The Stockholm Convention on Persistent Organic Pollutants announced a global ban on the manufacture and use of Endosulfan. India was not part of the ban.
2011	The Supreme Court banned the production, storage, sale, and use of Endosulfan in India.

The idea of banning Endosulfan in India received a couple of unfavorable responses from key agencies stating that the issue with Endosulfan was with its unscientific use rather than harmful health impacts. However, the Supreme Court of India imposed initially a temporary ban on the production, storage, sale and use of Endosulfan on May 13, 2011. Table 1 summarizes the developments of on regulation of Endosulfan in India.

1.2. Endosulfan and Its Different Trade Names

Table 2. Trade names of Endosulfan in different countries

Country	Trade names of Endosulfan
Bangladesh	Thiodan
Brunei	Thiodan, Fezdion
Chile	Parmazol E, Flavylon, Galgofan, Thiodan, Thionex, Thionyl methofan
India	Agrosulphan, AgiroSulphan, BanejSulphan, CiloSulphan, Endo Sulphan, E-Sulphan, Endo Chithin, Endocid, Endonit, Endomil, Endosol, Endostar, Endosun, Endotaf, Endostan, Endocing, Endocide, Endosulpher, Gaydan, GilnoreEndorifan, Hexa-sulphan, Hildan, Hockey Endosulfan, Hy-sulphan, KemuSulphan, Hilexute-Sulphan, KrushiEndosulfan, LusuSulphan, Marvel-Micosulphan, MicoThansulphan, ParySulphan, Pesticel, Remisfan, Sicosulphan, Solesulphan, Sujadin, Sulphan, TejSulphan, Thiodon, Thiokill, Thionel, Thionex, Thioton, Veg-fruThiotox, Veg-fruThiotex, Vikasulfan.
Indonesia	Thiodan, Fanodan, Dekasulfan.
Korea	Malix, Thiolix.
Pakistan	Siagon, Thiodan, Thioluxan.
Philippines	Atlas Endosulfan, Endosulfan, Contra, Endox, Thiodan.
Sri Lanka	Thiodan, Thionex, Endomack, Endocel, BaursEndosulfan, HarcrosHarcosan, Red Star Anglo-sulphan.
Thailand	Thiodan, A. B. Fan, Aggrodan, Agridan, Bensodan, Bensocarb, Beosit, Brook, Clement, Dew Dan, Dior 35, Dori, Dumpersan, E C Sulphan, Egodan, Endan, Endodan, Endosulfan, Endrew, Endye, Endyne, Etonic, Exxo-Z, Famcodan, Fortune, Freedan, Gardner, Gycin, Hor Mush, Hydrodan, J-teedan, Jack Dum, Kasidan, L P dan, Lordjim, Malix, Manyoo, Metrodan, Nayam, Newcodan, Nockdyne, Ox Xa, Patodan, Pestdye, Pro-d-dan, Sandan, Shevanex, Simadan, Sonydan, Summer, Tanadan, Teophos, Thanyacarb, Thimul, Thiofor, Torpidan, Urofen, Wephos, Zumic.
Other Names	Chlorthiepin, Cyclodan, Endox, Thifor, Thiomul, Thionate.

Despite the ban of Endosulfan in India in 2011, one of the key challenges is the availability of Endosulfan under different trade names in the market, and their usages in tea-producing estates. Endosulfan is available as formulations of an emulsifiable concentrate, wettable powder, ultra-low volume liquid, granules, dust and smoke tablets [5].

Endosulfan is sold with a number of trade names in different countries. The trade name, Thiodan is a commonly occurring pesticide used in tea-producing estates. The different trade names of Endosulfan sold in the market is presented in Table 2.

2. METHODOLOGY

Previous studies on Endosulfan concentration in the atmosphere, surface soil, surface water, and groundwater have been put together from various states of India. The Soxhlet extraction method was used for the extraction of the air samples [6-8]. Water samples were extracted using liquid-liquid extraction techniques except in the samples collected from the Hindon River [9] where solid phase extraction was employed. Soil samples were subjected to Soxhlet extraction except for the study from Thiruvallur [10] and Hisar, where accelerated solvent extractor [11] and flourosil column [12] was used. All extracts were cleaned using a silica-alumina column. Air samples were analyzed in Gas Chromatography-Mass Spectrometry (GC-MS) and Gas Chromatography Electron Capture Detector (GC-ECD). Water samples were mostly analyzed in GC-ECD with the exception of a study in Brahmaputra and Hooghly River [13] where water samples were analyzed in GC-MS. Soil samples were analyzed in GC-ECD but GC-MS was used for soil samples from seven cities of India [14].

2.1. Eco-Toxicological Risk Assessment

The ecotoxicological risk for river waters was assessed using a hazard quotient approach according to USEPA 1998 guidelines as estimated in another study [13]. The formula used for the calculation is given below:

$HQ = MEC/PNEC$

where HQ is the hazard quotient, MEC is the maximum reported environmental concentration i.e., the maximum concentration found in the river water and PNEC is the predicted no-effect concentration of aquatic species. The PNEC values of species under each organism group are taken from an earlier study on organochlorine pesticides in surface waters [13]. For all the rivers, the maximum concentration has been used for the calculation, except for the Yamuna River where we have used the mean value as the maximum value was not available.

3. Occurrences of Endosulfan Residues in India

3.1. Surface Water and Groundwater

It was observed that river water samples were more polluted with α-Endosulfan compared with β-Endosulfan [9, 15-17]. Only in the Hooghly and the Brahmaputra rivers, the dominance of β-Endosulfan was observed due to past usage of this pesticide in eastern and north-eastern states [13]. Higher concentration of endosulfan sulfate was observed in the Gomti River [18]. In the northern plains, α-Endosulfan isomers and presence of Endosulfan sulfate in the river water indicate a mixture of past and current usage of Endosulfan in the region [9, 11, 16, 19, 20]. Even after the strict ban, Endosulfan is still being used extensively by the local farmers under different trade names [17]. In the eastern part of the country, β-Endosulfan was more dominant [13]. Only in the sites near tea estate of the north-eastern region, α-Endosulfan and β-Endosulfan were detected. Endosulfan has been said to be heavily used in the tea estates [21]. Limited studies have been conducted on the rivers in the southern part of India. In the Kaveri river, residues of Endosulfan has been observed in the recent past [22].

Unlike surface water where α-Endosulfan was dominant in groundwater samples from across the country and showed even distribution of isomers and metabolites. Accidental spills, spray drift, soil texture, and run-in are suspected to play a vital role in Endosulfan contamination in groundwater [10]. The permeability of the geological layers of the soil also contributed to maximum contamination of pesticide residues in groundwater. β-Endosulfan and Endosulfan sulfate are more persistent in nature. They are mostly associated with soil contamination, over time the contaminants might penetrate to the groundwater source and pollute it.

3.2. Surface Soil

Overall, endosulfan sulfate dominated the organochlorine pesticides (OCP) level in Indian soil but a relatively higher level of α-Endosulfan was observed in the northern soil [11, 12, 23]. Predominant levels of Endosulfan sulfate were also observed in metropolitan cities [14]. This reflects past usage of the technical mixtures.

Endosulfan levels in southern India were attributed to technical Endosulfan usage for cashew plantations and higher α-Endosulfan level in Goa and Mumbai is suspected from the ongoing usage of Endosulfan mainly for cotton cultivation practiced in the western and central parts of India [14]. In the southern part of India, soils with high organic carbon were having higher endosulfan sulfate residues indicating that high organic carbon might play a vital role in the persistence of pesticide residues [24]. In the eastern part of India endosulfan sulfate was dominant followed by α-Endosulfan which is linked with the usage of Endosulfan in agricultural purposes [14, 25, 26].

3.3. Air

Although there was the dominance of endosulfan sulfate in major Indian cities such as New Delhi, Chennai, and Kolkata but cities like Mumbai which is closer to cotton cultivation in the western part of India where Endosulfan has been extensively used, showed the dominance of β-Endosulfan [6]. In the north-eastern part of India, studies in Manipur showed a higher level of α-Endosulfan during hot and rainy seasons.

These may be from the surrounding cultivable lands and atmospheric transport from nearby tea estates [8]. In northern parts of the country, Endosulfan was fairly uniform and slightly higher at the sites in agricultural regions. But this trend was different from agricultural sites from southern India [7].

3.4. Eco-Toxicological Risk Assessment Discussion

It can be observed from Figure 1 that water from the Ganga river in Varanasi and Hindon river show an adverse impact on all the organisms since the mean hazard quotient (HQ) is higher than 1. Lower trophic organisms like phytoplankton and zooplankton were less affected by Endosulfan pollution in the river water but the concern arises when we consider the fish species. Excluding the Hooghly river, the mean value of HQ is higher than 1 for fishes of all the rivers. Out of the five fish species considered, three are edible fish species like Rohu and Catla, which were also at risk.

Higher eco-toxicological impacts predicted for edible fish is very important considering the possibility of fish impacting humans through dietary intake particularly in eastern and north-eastern India where these fish are commonly consumed. Indian standards for drinking water have a permissible limit of 0.4 µg/L for Endosulfan [27]. Groundwater from many locations in New Delhi [28], Kasargod [29] and Thiruvalluvar [10] had Endosulfan content more than the permissible limit and hence not fit for drinking purposes.

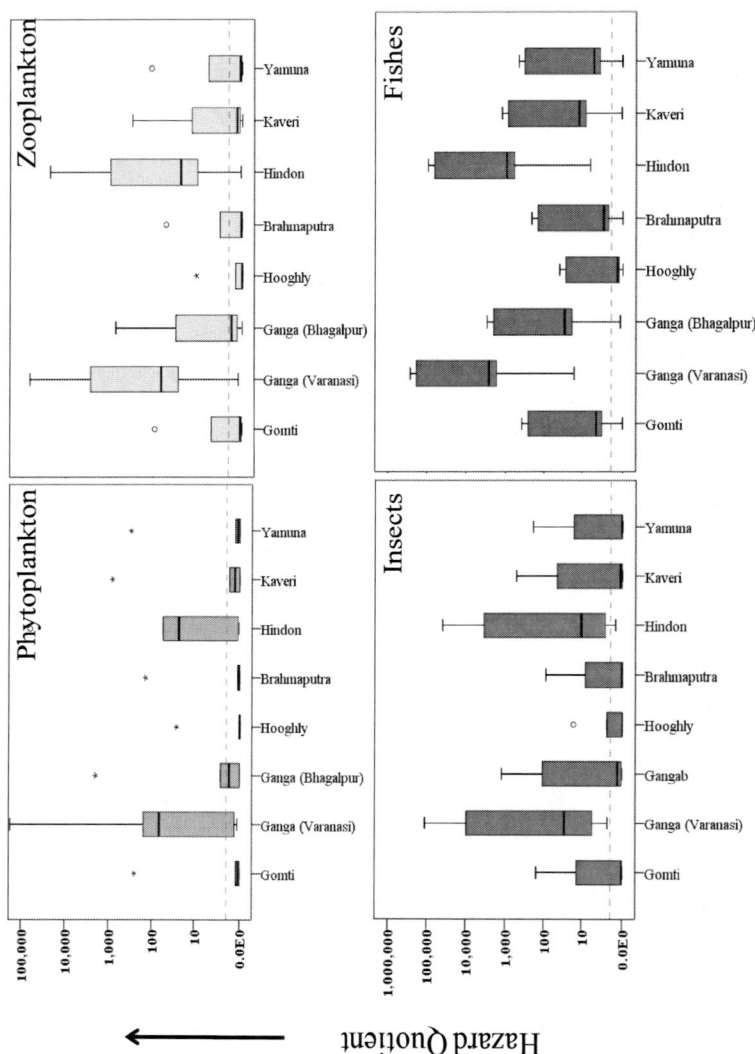

Figure 1. Ecotoxicological risk assessment of endosulfan in surface water of Indian rivers. Note: Values were taken from previous studies in Gomti river [18], Ganga (Varanasi) [16], Ganga (Bhagalpur) [17], Hooghly and Brahmaputra [13], Hindon [9], Kaveri [22], Yamuna [15].

3.4.1. A Case Study of Endosulfan Poisoning in Kerala, India

Endosulfan was sprayed aerially over the cashew plantations in Kasaragod district in Kerala, India for around three decades [30]. The usage of Endosulfan started in 1978 when they were widely used to control the tea mosquito bug (*Helopeltistheivora*) in cashew plantations. It was applied 3 times a year over an area covering 15 Gram Panchayats (local government at the village level) in Kasaragod. Several symptoms of mental and physical disorders (especially in children) were reported in Kasargod that was linked to over 20 years of aerial spraying of Endosulfan over cashew nut plantations. Studies in Kasargod area showed that the sediment and water of 12 streams (used by villagers) originating in the plantation were heavily contaminated with Endosulfan even outside the spray season [31]. The study conducted on various environmental as well as human samples such as soil, water, blood, milk etc in Padre village in Kasargod [32] showed very high residue of Endosulfan compared with the maximum permissible level [3]. The villagers were also directly exposed to spray drift that resulted in long-term congenital, reproductive, and neurological symptoms and ailments. About 197 deteriorating health cases were reported from just 123 households [3]. Chronic morbidity was 70 percent higher than the normal. There were some severe observations of similar effects in animals too.

Higher prevalence of neurobehavioral disorders, congenital malformations in females, as well as abnormalities related to the male reproductive system was observed by Endosulfan exposure in Kasargod [33]. In August 2001, the Government of Kerala imposed a complete ban on Endosulfan in all crops and plantations in the state. After almost a decade, in 2010, a comparative study still showed health impacts like reproductive morbidity, sexual maturity congenital anomalies and cancer in younger ages among the people residing in Kasargod due to Endosulfan [34]. Similar health impacts were observed in the neighboring Belthangady taluk in Dakshin Kannada district of Karnataka, again due to aerial spraying of Endosulfan in cashew plantations. In 2011, Endosulfan was banned in Karnataka.

At the global level, due to its high toxicity and persistence that brought about various health and environmental hazards, Endosulfan was phased out from many countries and was accepted as Persistent Organic Pollutant (POP). However, Endosulfan was allowed to be used in India until the Supreme Court ordered a complete ban on Endosulfan on 13 May 2011.

4. Sustainable Practices That Eliminate the Use of Endosulfan

4.1. Alternatives to Endosulfan

A number of chemical and non-chemical alternatives to Endosulfan have been applied and proved to be successful in some regions in developed and developing countries. Some of these alternatives are being applied in countries where Endosulfan has been banned or is being phased out.

Some countries prefer to use Endosulfan in pollination management, integrated pest management systems, and because of its "generic" nature, consider it to be effective against a broad range of pests. A few others want to continue to use Endosulfan to allow time to adapt and phase-in of alternatives. Some of the chemical, biological and semi-chemical alternatives to Endosulfan researched and used in India are summarized in Tables 3-5.

Some states of India have successfully adopted and followed the non-chemical alternative to Endosulfan. These Indian states include Andhra Pradesh (non-pesticidal management, NPM), parts of Punjab, Maharashtra, Karnataka (zero budget farming) and Sikkim (organic farming). These practices have helped them to bring economic stability among farmers, also preserved the environment and brought back the pollinators.

Table 3. List of chemical, biological and semio-chemical alternatives of Endosulfan in India [41]

Chemical alternatives (including plant extracts)

Alternative plant protection products	Crop Type	Pest Type
Cypermethrin	Paddy	Leaf folder
Lampda cyhalothrin	Paddy	Leaf folder
Quinalphos	Paddy	Hispa/case worm/cut worm Swarming caterpillar/ *surti*caterpillar
	Sorghum	Defoliators
	Jute	Semilooper
	Sugarcane	Top shoot borer; Internode bore
Monocrotophos	Paddy	Hispa/case worm/cut worm Swarming caterpillar/ *surti* caterpillar
	Green gram, black gram	Pod borer
	Groundnut	Leaf miner
	Niger	Lucern caterpillar defoliator
	Soyabean	Stem fly defoliator
	Jute	Bihar hairy caterpillar
	Indigo	Caterpillar
	Mango	Mealybug
	Guava	Bark eating caterpillar
Chlorpyrifos	Groundnut	*Helicoverpa sp./Spodoptera* sp./other leaf-eating caterpillars
	Potato	Cutworm
Carbaryl	Paddy	Swarming caterpillar/ *surti* caterpillar
	Linseed	Defoliator
	Maize	Corn earworm/defoliator
	Sugar cane	Top shoot borer; Internode borer
	Lady Finger *(Bhindi)*	Leaf roller
	Curcubits	Red pumpkin
	Cabbage/Cauliflower	Cabbage borer, Tobacco caterpillar, Cabbage butterfly
	Pea	Pod borer
	Mango	Mango hopper
	Guava	Cater capsule borer
Triazophos	Cotton	Spotted bollworm; pink bollworm; *Helicoverpa sp.*
	Pigeon pea *(Arhar)*	Pod borer
	Green gram, black gram	Pod borer
	Groundnut	*Helicoverpasp./Spodopterasp* /other leaf-eating caterpillars
	Soyabean	Leaf roller Leaf miner

Table 3. (Continued)

Alternative plant protection products	Crop Type	*Pest Type*
Triazophos	Soyabean	Leaf roller Leaf miner
	Chilli	Fruit borer
	Tomato	Fruit borer (*Helicoverpaarmigera*)
	Pea	Pod borer
Acephate	*Arhar* [pigeon pea]	Pod borer Defoliators
	Cabbage/Cauliflower	Leaf Webber (*Crocidolomabinotalis*)
Methyl oxydemeton	*Arhar* [pigeon pea]	Pod bug
	Safflower	Bihar hairy caterpillar
Imidacloprid	*Arhar* [pigeon pea]	Pod bug
Ethofenprox*	Mustard	Leaf Webber
Dichlorvos*	Sunflower	Cutworm
Dimethoate*	Safflower	Bihar hairy caterpillar
	Mesta	Jassid
Phosalone*	Soyabean	Leaf roller, Leaf miner, Stem fly defoliator
	Jute	Semilooper
Carbofuran*	Maize	Stalk Borer
	Sugarcane	Top shoot borer; Internode borer
Fenvalerate*	Sorghum	Defoliators a;
Spinosad*	Cotton	Red cotton bug; Dusky cotton bug
Indoxacarb*	Cotton	Red cotton bug; Dusky cotton bug
Dicofol*	Jute	Mites
	Chilli	Fruit borer
Propargite*	Jute	Mites
NKSE*	Citrus	Lemon butterfly
Monocrotophos	Pigeon Pea *(Arhar)*	Pod fly (*Melanagromyzaobtusa*)
Malathion, Quinalphos, Methyl parathion Carbaryl	Mustard	Sawfly (*Athalialugensproxima*)
Quinalphos, Phosalone, Malathion, Deltamethrin, Cypermethrin	Sesamum	*Antigastra sp.*/ Pod capsule borer
Quinalphos, Triazaphos, Methyl parathion, Malathion	*Ragi (Eleusinecoracana)*,	Millet shoot fly
Phosalone, Malathion, DAS (specific active substance could not be identified)	Sesamum	Hawk moth (Sphinx caterpillar),
Neem (Azadirachtin) (plant extract)*	Sunflower	Castor semilooper (*Achaea janata*)
	Mustard	leaf and pod caterpillar

Alternative plant protection products	Crop Type	*Pest Type*
Neem seed kernel suspension (plant extract)*	Sunflower	Castor semilooper (*Achaea janata*)
Neem based pesticide (plant extract), Imidacloprid, Spinosad, Acetamiprid, Buprofezin, Novaluron, Indoxacarb, Flubendiamide, Thiomethoxam, Emamectinbenzoate, Chlorantraniloprole	Not Specified	Not Specified
Neem(plant extract)*	Groundnut	*Helicoverpasp./Spodopterasp./*other leaf-eating caterpillars
	Sunflower	Defoliators
	Bengal gram	Pod borer

Table 4. Biological control alternatives

Alternative Biological Control Agent	Crop type	Pest Type
Parasitic wasp *Trichogramma sp.*	Cotton	Not specified
Conserve *Ormyrus sp.* (parasite of pod fly)	Arhar (pigeon pea)	Pod fly (*Melanagromyzaobtusa*)
Conserve *Perilissuscingulator* (parasites the grubs); bacterium *Serratia marcescens*	Mustard	Sawfly (*Athalialugensproxima*(Klug)
Braconhebator; *B. Brevicornis*; *Phanerotomahandecasisella*	Sesamum	Shoot Webber
Cantheconidiafurcellata; *Cicindellaspp*; Parasitoids*Tratnalaflavoorbitallis*; Parasitoids*Campoplex sp.*; Parasitoids*Erioborus sp.*; Parasitoids*Temeluchabiguttula*; Parasitoids*Apanteles spp.*; Parasitoids*Cremastusflavoorbitalis*	Sesamum	*Antigastra sp.*/ Pod capsule Borer
egg parasite *Anastatusacherontiae*; larval parasite *Sarcophaga sp.*; larval parasite *Zygobothria ciliate Walp*; larval parasite *Apantelesacherontiae*;	Sesamum	Hawk moth (Sphinx caterpillar)
*Bacillus thuringiensis**	Mustard	leaf and pod caterpillar
*Trichogramma sp.**; *Apanteles sp.**; *Bracon sp.**; *Chrysopa sp.**; Ladybird beetles*; *Bacillus thuringiensis**	Sunflower	Castor semilooper (*Achaea janata*)
*Telenomus dingus**; *Trichogramma sp.**; fungus *Beauvariabassiana**	Ragi (*Eleusinecoracana*)	Pink borer
*Trichogrammcchilonis**	Paddy	Leaf folder, Hispa/case worm/cutworm

Table 4. (Continued)

Alternative Biological Control Agent	Crop type	Pest Type
Helicoverpaarmigera nuclear polyhedrosis virus (NPV)*	*Arhar* [pigeon pea]	Pod borer
*Bacillus thuringiensis**	*Arhar* [pigeon pea]	Pod borer
Helicoverpaarmigera nuclear polyhedrosis virus (NPV)*; *Bacillus thuringiensis**	Bengal gram	Pod borer
Helicoverpaarmigera nuclear polyhedrosis virus (NPV)*; *Bacillus thuringiensis**	Green gram, black gram	Pod borer
Spodopteralitura NPV*	Groundnut	Defoliator (*Spodopteralitura*)
Helicoverpaarmigera nuclear polyhedrosis virus (NPV)*	Groundnut	*Helicoverpa*/*Spodoptera*/other leaf-eating caterpillars
*Bacillus thuringiensis**	Mustard	Leaf Webber
Helicoverpaarmigera nuclear polyhedrosis virus (NPV)*	Sunflower	*Helicoverpasp.* (head borer)
Helicoverpaarmigera nuclear polyhedrosis virus (NPV)*; *Bacillus thuringiensis**	Sorghum	Gram pod borer
*Trichogrammachilonis**; *Bacillus thuringiensis**; *Helicoverpaarmigera* (NPV)*	Cotton	Spotted bollworm; pink bollworm; *Helicoverpa*; Red cotton bug; Dusky cotton bug
*Trichogramma japonicum**; *Trichogrammachilonis**	Sugarcane	Top shoot borer; Internode borer
*Trichogrammachilonis**; *Bacillus thuringiensis**; *Helicoverpaarmigera* (NPV)*	Tomato	Fruit borer (*Helicoverpaarmigera*)
*Bacillus thuringiensis**	Cabbage/ Cauliflower	Cabbage borer; Leaf Webber (*Crocidolomabinotalis*); Cabbage Butterfly
*Bacillus thuringiensis**; *Helicoverpaarmigera* (NPV)*	Pea	Pod borer

Table 5. Semi-chemical alternatives

Alternative semi chemical	Crop type	Pest Type
pheromone	Groundnut	Defoliator (*Spodopteralitura*)
pheromone	Sunflower	*Helicoverpasp.* (head borer)
pheromone	Tomato	Fruit borer (*Helicoverpaarmigera*)

4.2. Initiatives to Control Endosulfan Usage in India

4.2.1. Non-pesticidal management (NPM)

The NPM program in the southern Indian state of Andhra Pradesh is practiced in 3000 villages over an area of 1.7 million acres and aims to employ safer sustainable alternatives to Endosulfan that would eliminate the need for pesticides [35]. NPM shows that pests can be managed effectively by locally grown resources and with timely action. Community-based organizations backed by government organizations can create wonders for the farming community by way of supporting them to practice NPM. The system of NPM aims at using 'zero pesticides' and has adopted different practices as follows [35].

- To attract moths, light traps and bonfires are used by farmers.
- To eliminate the insects that suck sap of plants, yellow and white sticky boards are placed in the field to attract and kill the insects.
- To eliminate and kill field pests, pheromone traps are set up.

A very effective and popular method of NPM is using biological pesticides such as neem seed-kernel extracts and chili-garlic extracts to control bollworms and sucking insects. The farmers also use other local plants to make biological pesticides. The farmers make an extract using cow dung and urine, which is used both for pest control (to control aphids and leafhoppers) and as a fertilizer. The leaves on which insect eggs are laid are removed by hand. Insects are likely to lay their eggs on plants like castor and marigold, which are often termed as 'trap crops' that can be picked off and removed easily.

4.2.2. Zero Budget Natural Farming (ZBNF)
ZBNF involves growing of crops naturally without using fertilizers or pesticides, and focusing on 'zero budget' or zero net cost of production of crops. It requires minimal, low-cost and locally available inputs for seed treatments and other inoculations such as cow-dung and cow urine.

Adoption of ZBNF has shown benefits such as increased income of farmers, better crop yield, soil fertility, seed diversity, improved quality of produce, etc. ZBNF is a set of sustainable farming methods that was started from the southern Indian state of Karnataka and has now spread to other Indian states. Around 100,000 farmers practice ZBNF in Karnataka. ZBNF runs with the voluntary involvement of farmer families and other community members. To run the program, formal organizations are not required to be involved and it does not require a budget [37]. Mr. Subhash Palekar, the former agricultural scientist, is referred to as the father of the ZBNF movement in the state of Karnataka, India. He has developed the basic "Toolkit" of ZBNF methods, which include the following four methods:

- *Jivamrita/Jeevamrutha* is a fermented microbial culture that provides soil nutrients and acts as a catalytic agent that increases activities of earthworm in the soil.
- *Bijamrita/Beejamrutha* is a treatment used for seeds, seedlings and protects young roots from soil-borne diseases.
- *Acchadana*/Mulching includes:
  a) Soil mulch to protect topsoil during cultivation and does not destroy it by tilling;
  b) Straw mulch that is the dried biomass waste of previous crops;
  c) Live mulch (symbiotic intercrops and mixed crops).

4.2.3. Organic Farming

Organic farming is another important sustainable agriculture element in developing countries. It includes a mix of low external input technology, environmental conservation, and input/output efficiency. It also provides access to good price markets to farmers (through certifications). Since it is deemed to be an effective method of improving productivity and food security, non-governmental organizations and farmers' groups are adopting this system [37]. Sikkim, one of the north-eastern State of India, has successfully implemented a policy to switch all of its agriculture into organic production, by way of eliminating all forms of chemicals from agricultural practices and using options like enriched rural compost, vermicomposting, bio-fertilizers, green manure and organic amendments/fertilizers (dolomite and rock phosphates) [37]. In 2016, Sikkim became India's first 100 percent organic State. In India, nine other States Karnataka, Mizoram, Kerala, Andhra Pradesh, Himachal Pradesh, Madhya Pradesh, Tamil Nadu, Maharashtra, and Gujarat have adopted an organic farming policy or law.

4.2.4. Integrated Pest Management (IPM)

IPM is a broad-based approach that integrates practices for controlling pests. It considers all available pest control techniques and subsequent integration of best practices/measures that reduces or eliminates pests and keeps the pesticide use at levels that are viable from the economic, human health and environmental points of view. National Research Centre for Integrated Pest Management (NCIPM), India was established in 1988 to adopt IPM technologies for reducing production costs and minimizing environmental and public health hazards. In India, the use of IPM has been successful for field (rice and pulses), commercial (cotton and oilseeds), vegetable (cauliflower and cabbage) and fruit (kinnow and khasi mandarins) crops, development of light traps and implementation of area-wide ICT-based e-pest surveillance, and awareness generation amongst stakeholders including farmers [39].

5. THE ROAD AHEAD: KEY ACTION POINTS AND RECOMMENDATIONS

5.1. Key Action Points

A number of legislations are available in India that aims to protect human health and the environment (directly or indirectly) from the harmful effects of Endosulfan. These legislations include the following [39] :

- Article 21 of the Constitution of India;
- The Directive Principles of State Policy, Article 48A, Article 51A;

- The Environment (Protection) Act & Rules, 1986;
- The Insecticides Act, 1968;
- The Prevention of Food Adulteration Rules, 1955;
- The Consumer Protection Act, 1986.

Beside the Stockholm Convention, India is a signatory to the Strategic Approach to International Chemicals Management (SAICM), which cover various aspects of management, such as use, manufacture, transport of harmful chemicals like Endosulfan and ensure the protection of the environment and human health.

Although various regulations that ban or control the use of Endosulfan are in place, critical levels are found in the environment (especially in the aquatic environment) and were found to have long-lasting effects on human health, affecting women and children. This strongly indicates that the current regulations in India have not been fully implemented to curb the ill-effects of Endosulfan. It is important to note that future policies to manage Endosulfan in India have to consider the present and past aspects and work in an integrated manner that would be directed towards a safe and sustainable future.

It is time the Indian government takes a strong stand to eliminate and eventually comply with a complete ban on Endosulfan from all agricultural practices and systems.

5.2. Policy Recommendations

5.2.1. For the Government

Unlike other developed countries, there is a lack of a rigorous system to review registered and banned pesticides in India, which needs to be done periodically to ensure enforcement of national and state level regulations. There is a lackadaisical attitude amongst state government and state agriculture authorities on enforcement of the ban on Endosulfan. In 2011, it was reported that despite its ban in Kerala, Endosulfan trade was going on by way of entering through the neighboring state of Tamil Nadu, and

Endosulfan was being illegally dumped by manufacturers that seeped into the nearby water bodies, causing a human health hazard. It was also reported that in states like Punjab and Odisha, Endosulfan was being sold in shops, either secretly or openly. It needs to mention that Punjab imposed a state ban on Endosulfan only in February 2018.

In 2016, the Government of India appointed an Expert committee under Dr. Anupam Verma of Indian Agriculture Research Institute (IARI), to carry out a technical review of 66 pesticides that are banned, restricted, or continued to be registered in India. The Verma panel, however, did not review the status of Endosulfan ban, as it was a sub-judice matter at that time. It may be reiterated that the protection of public health and the environment is the key responsibility of the government. It is time the government agencies stop paying importance to the profit-making needs of business organizations and take appropriate measures to protect public health and the environment. In this regard, empowerment of officials of concerned government departments through training and capacity building programs may be considered.

Government agencies and institutions should proactively explore and use all the possibilities and opportunities available under various international treaties such as the Stockholm Convention and the Basel Convention. The Basel Convention controls transboundary movement (imports, exports and transits) of POPs wastes and its disposal by the provisions of the convention as well as by the provisions of national law.

All technical, advisory and financial aspects should be well considered and applied towards eliminating and ban of Endosulfan, and shifting focus towards safer and sustainable alternatives. The Government should proactively promote non-chemical alternatives to Endosulfan as well make regulatory amendments for the adoption of initiatives to controlled use of Endosulfan like organic farming, zero budget natural farming and integrated pest management.

5.2.2. For Farmer Stakeholder Groups

The inclusion of farmers in Endosulfan management is imperative. Farmers are the end users, and they handle large quantities of pesticides.

Therefore, it is important to ensure the protection of their health and those of their families. The Agriculture Department relies on and works with agricultural agencies and institutions to provide technical, regulatory and support services. These agencies may deal directly with farmer representatives, who in turn, can strongly influence and help the implementation of any program that manages and regulates Endosulfan use or adopt safer alternatives. The government agencies often constitute committees related to new pesticide registration, etc. Farmer representatives should be included in such committees.

For effective management of pesticides, it is also important to impart training and conduct capacity building programs for farmers. Such programs should aim at:

- Safe storage, handling, and disposal of pesticides at their level (users).
- Emphasize on mandatory use of protective gears such as masks, protective gloves and clothing while working.
- Provide required demonstrations to farmers on optimum quantity for usage of pesticides.
- Encourage usage of non-chemical means of pest control.

5.2.3. For the Scientific Community

A scientific research-based recommendation on remedial and preventive actions to control Endosulfan, and suggestions for alternate uses, is needed at present. It is imperative for scientists to take up research and development on bio-pesticides and explore green alternatives to Endosulfan. It is equally crucial for the scientific community to translate and apply such research results in real-world settings. This is possible through the establishment of strong linkages between scientists and other decision makers (such as government, industry, and NGOs), and subsequent sharing and dissemination of the research outcomes in an effective manner. In that way, the scientific community can play an active role in setting the agenda for management of Endosulfan and protection of public health at large.

5.2.4. *For the NGOs/CSOs*

Non-government organizations (NGOs) and Civil Society Organizations (CSOs) is a crucial institutional link that can provide effective support where the government and private players are falling short. NGOs are very prominent in effective implementation of government programs and in influencing awareness programs. For example, Self-Help Groups (SHGs) and activist organizations can be engaged in curbing misuse of Endosulfan. Also, NGOs can play an active role in training the farmers on practices of safe handling of Endosulfan; conducting mass awareness campaigns on organic farming, adopting safer alternatives of Endosulfan, etc. NGOs can also make meaningful contributions to reporting of non-compliance to appropriate authorities because they have direct connection with communities (farmers). For improving access to remote areas, the use of services of NGOs to engage communities can be very useful that are otherwise difficult to manage through the formal sector routes. Although a very strong NGO network exists in India, the influence of NGOs can be enhanced through further recognition (by government) of their contribution, improved support (of technical, financial, training and policy aspects) in a structured and planned manner, and strengthen coordination among the stakeholders such as farmers, scientists and policymakers.

ACKNOWLEDGEMENT

This work was supported by Selective Excellence initiative of SRM Institute of Science and Technology under signature programs competition, criteria for academic excellence, SRM Institute of Science and Technology Kattankulathur campus, and Research Excellence Inititative of Mu Gamma Consultants. Sanjenbam Nirmala Khuman was supported by the Council of Scientific & Industrial Research Ministry of Science & Technology, Govt. of India for Senior Research Fellowship (file no:09/1045(0020)2K18 EMR-1).

REFERENCES

[1] Health UDo, Services H. *Toxicological profile for Endosulfan.* Agency for toxic substance and disease registry, Atlanta. 1990.

[2] Joshi SSMaS. *Tracking decades-long Endosulfan tragedy in Kerala* available at https://www.downtoearth.org.in/coverage/health/tracking-decades-long-endosulfan-tragedy-in-kerala-56788. 2018.

[3] Watts M. *Endosulfan* available at http://www.pan-germany.org/download/Endo_09_PANAP_monograph_2nd%20Edition.pdf 2009.

[4] Yadav IC, Devi NL, Syed JH, Cheng Z, Li J, Zhang G, et al. Current status of persistent organic pesticides residues in air, water, and soil, and their possible effect on neighboring countries: A comprehensive review of India. *Science of the Total Environment.* 2015; 511:123-37.

[5] Organization FaA. *Non Pesticide Management in Andhra Pradesh, India* available at http://www.fao.org/fileadmin/templates/esw/esw_new/documents/SARD/good_practices_asia/4_pest_management_India.pdf2005.

[6] Chakraborty P, Zhang G, Li J, Xu Y, Liu X, Tanabe S, et al. Selected organochlorine pesticides in the atmosphere of major Indian cities: levels, regional versus local variations, and sources. *Environmental science & technology.* 2010;44(21):8038-43.

[7] Pozo K, Harner T, Lee SC, Sinha RK, Sengupta B, Loewen M, et al. Assessing seasonal and spatial trends of persistent organic pollutants (POPs) in Indian agricultural regions using PUF disk passive air samplers. *Environmental Pollution.* 2011;159(2):646-53.

[8] Devi NL, Qi S, Chakraborty P, Zhang G, Yadav IC. Passive air sampling of organochlorine pesticides in a northeastern state of India, Manipur. *Journal of Environmental Sciences (China).* 2011; 23(5):808-15.

[9] Ali I, Singh P, Rawat M, Badoni A. Analysis of organochlorine pesticides in the Hindon river water, India. *Journal of Environmental Protection Science.* 2008;2:47-53.

[10] Jayashree R, Vasudevan N. Organochlorine pesticide residues in groundwater of Thiruvallur district, India. *Environmental monitoring and assessment.* 2007;128(1-3):209-15.

[11] Kumari B, Madan V, Kathpal T. Status of insecticide contamination of soil and water in Haryana, India. *Environmental Monitoring and Assessment.* 2008;136(1-3):239-44.

[12] Kumar B, Kumar S, Gaur R, Goel G, Mishra M, Singh SK, et al. Persistent organochlorine pesticides and polychlorinated biphenyls in intensive agricultural soils from North India. *Soil Water Res.* 2011;6(4):190-7.

[13] Chakraborty P, Khuman SN, Selvaraj S, Sampath S, Devi NL, Bang JJ, et al. Polychlorinated biphenyls and organochlorine pesticides in River Brahmaputra from the outer Himalayan Range and River Hooghly emptying into the Bay of Bengal: Occurrence, sources, and ecotoxicological risk assessment. *Environmental Pollution.* 2016;219:998-1006.

[14] Chakraborty P, Zhang G, Li J, Sivakumar A, Jones KC. Occurrence and sources of selected organochlorine pesticides in the soil of seven major Indian cities: Assessment of air-soil exchange. *Environmental pollution.* 2015;204:74-80.

[15] Aleem A, Malik A. Genotoxicity of the Yamuna river water at Okhla (Delhi), India. *Ecotoxicology and environmental safety.* 2005;61(3):404-12.

[16] Nayak A, Raha R, Das A. Organochlorine pesticide residues in middle stream of the Ganga River, India. *Bulletin of environmental contamination and toxicology.* 1995;54(1):68-75.

[17] Leena S, Choudhary S, Singh P. Pesticide concentration in water and sediment of River Ganga at selected sites in middle Ganga plain. *International journal of environmental sciences.* 2012;3(1):260.

[18] Malik A, Ojha P, Singh KP. Levels and distribution of persistent organochlorine pesticide residues in water and sediments of Gomti River (India) a tributary of the Ganges River. *Environmental Monitoring and Assessment.* 2009;148(1):421-35.

[19] Kaushik A, Sharma H, Jain S, Dawra J, Kaushik C. Pesticide pollution of river Ghaggar in Haryana, India. *Environmental monitoring and assessment.* 2010;160(1-4):61.
[20] Sankararamakrishnan N, Sharma AK, Sanghi R. Organochlorine and organophosphorus pesticide residues in groundwater and surface waters of Kanpur, Uttar Pradesh, India. *Environment International.* 2005;31(1):113-20.
[21] Gurusubramanian G, Rahman A, Sarmah M, Roy S, Bora S. Pesticide usage pattern in tea ecosystem, their retrospects and alternative measures. *Journal of Environmental Biology.* 2008;29(6):813-26.
[22] Rajendran RB, Subramanian A. Pesticide residues in water from river Kaveri, South India. *Chemistry and Ecology.* 1997;13(4):223-36.
[23] Singh KP, Malik A, Sinha S. Persistent organochlorine pesticide residues in soil and surface water of northern Indo-Gangetic alluvial plains. *Environmental monitoring and assessment.* 2007;125(1-3):147-55.
[24] Jayashree R, Vasudevan N. Residues of organochlorine pesticides in agricultural soils of Thiruvallur district, India. *International journal of food, agriculture, and the environment.* 2006;4(1):313-6.
[25] Devi NL, Chakraborty P, Shihua Q, Zhang G. Selected organochlorine pesticides (OCPs) in surface soils from three major states from the northeastern part of India. *Environmental monitoring and assessment.* 2013;185(8):6667-76.
[26] Devi NL, Yadav IC, Raha P, Shihua Q, Dan Y. Spatial distribution, source apportionment and ecological risk assessment of residual organochlorine pesticides (OCPs) in the Himalayas. *Environmental science and pollution research.* 2015;22(24):20154-66.
[27] Specification ISDW. IS 10500. (2012). *Bureau of Indian Standards.* 2012.
[28] Mukherjee I, Gopal M. Organochlorine insecticide residues in drinking and ground water in and around Delhi. *Environmental monitoring and assessment.* 2002;76(2):185-93.

[29] Akhil P, Sujatha C. Prevalence of organochlorine pesticide residues in groundwaters of Kasargod District, India. *Toxicological & Environmental Chemistry*. 2012;94(9):1718-25.

[30] Embrandiri A, Singh RP, Ibrahim HM, Khan AB. An epidemiological study on the health effects of Endosulfan spraying on cashew plantations in Kasaragod District, Kerala, India. *Asian journal of epidemiology*. 2012;5(1):22-31.

[31] Wani KA. *Handbook of Research on the Adverse Effects of Pesticide Pollution in Aquatic Ecosystems*: IGI Global; 2018.

[32] Environment CfSa. *A Centre for Science and Environment report on the contamination of Endosulfan in the villagers* available at http://www.indiaenvironmentportal.org.in/files/CSE_report.pdf. 2001.

[33] Dayakar MM, Shivprasad D, Dayakar A, Deepthi CA. Assessment of oral health status among Endosulfan victims in Endosulfan relief and remediation cell-A cross-sectional survey. *Journal of Indian Society of Periodontology*. 2015;19(6):709.

[34] Dept of community medicine GMCC. *Epidemiological studies related to health in Endosulfan affected areas at Kasargod District, Kerala* available at https://cdn.cseindia.org/userfiles/kasargod_district_report_page1-20.pdf. 2011.

[35] United Nations. *Case studies* available athttps://sustainabledevelopment.un.org/content/dsd/resources/res_pdfs/publications/sdt_toxichem/practices_sound_management_chemicals_case_ex_1-14.pdf. 2018.

[36] Food and Agriculture Organization F. *Zero Budget Natural Farming in India* available at http://www.fao.org/3/a-bl990e.pdf2018.

[37] FAO. *International Conference on Organic Agriculture and Food Security* available at http://www.fao.org/tempref/docrep/fao/meeting/012/ah953e.pdf. 2007

[38] Taneja S. *Sikkim is 100% organic! Take a second lo*ok available at https://www.downtoearth.org.in/news/agriculture/organic-trial-57517 2017.

[39] Vennila S, Ajanta B, Vikas K, Chattopadhyay C. *Success Stories of Integrated Pest Management in India.* available at http://www.ncipm.res.in/NCIPMPDFs/folders/Success%20stories.pdf. 2016.

[40] Radhika Yadav SM. *Banning Endosulfan in India ~ Consumers' health rights v. Consumers' economic rights*: National Law School of India University; 2012.

[41] Convention S. *Technical Endosulfan and its related isomers* available at http://chm.pops.int/Implementation/Alternatives/Alternativesto POPs/ChemicalslistedinAnnexA/TechnicalEndosulfan/tabid/5867/Default.aspx 2008.

Chapter 4

ENDOSULFAN EVALUATION IN HUMAN SAMPLES

Virgínia Cruz Fernandes, Maria Luísa Correia-Sá, Maria Luz Maia, Cristina Delerue-Matos and Valentina Fernandes Domingues[*]

REQUIMTE/LAQV, Instituto Superior de Engenharia do Porto, Instituto Politécnico do Porto, Porto, Portugal

ABSTRACT

Endosulfan,(1,4,5,6,7,7-hexachloro-8,9,10-trinorborn-5-en-2,3-ylenebismethylene)sulfite, is a cyclodiene and a popular agricultural insecticide which in its pure form comprises a mixture of two stereoisomers, α and β. Endosulfan is highly persistent in the environment because of their resistance to chemical and biological degradation. In addition, their solubility in lipids contributes to their bioaccumulation and biomagnification through food chains, increasing the potential risk to human health. Exposure to hazardous pesticides is of great concern to the general population, because of the widespread use of the compounds in agriculture, public health, home gardening, and industry. The best way to

---
[*] Corresponding Author's Email: vfd@isep.ipp.pt.

measure the exposure of pesticides (or their metabolites) is to measure them directly in biological fluids (human biomonitoring). Reliable analysis of residual endosulfan in human samples usually involves detection of endosulfan α, endosulfan β, and their metabolites (sulfate, lactone, ether, and diol). The development of fast, sensitive, reliable and accurate analytical procedures for the determination of endosulfan and/or metabolites at low levels. Gas Chromatography (GC) with electron capture detector (ECD) and mass spectrometry (MS) have been the most widely used analytical techniques for determining low parts-per-billion (ppb) to parts-per-trillion (ppt) levels of endosulfan (α and β) and their metabolites. In this chapter, relevant issues, namely, sample preparation, analytical techniques employed, and the levels endosulfan in reported in different biomaterials are discussed.

Keywords: endosulfan, human samples, sample preparation, gas chromatography

1. INTRODUCTION

Pesticides are substances used to prevent, destroy, repel, or mitigate any pests, ranging from insects, animals, and weeds to microorganisms such as fungi, molds, bacteria, viruses, and other organisms that compete with humans for food, destroy property, spread disease, or are considered a nuisance [1]. The use of synthetic pesticides for agricultural and other purposes began 50 years ago. A number of studies have reported the human exposure of endosulfan in a number of complex biological samples, namely blood/serum/plasma, umbilical cord blood, adipose tissue, breast milk, urine, human tissues, semen, follicular fluid and placenta [2-4]. As a result of their extensive use, exposure to hazardous pesticides is a concern to the general population [5]. Chronic toxicity can result from occupational exposure or common consumption of contaminated food and drinking water. A number of studies have shown the ability of pesticides accumulation in various environmental media (air, water, sediments, soil) [6-8] and food products [9-11]. In order to assess the health risk of these substances and their metabolites, constant monitoring of endosulfan in the different biological specimen is required [5]. Among pesticides, organochlorine pesticides (OCPs) are a widely used class of insecticides. Organochlorine pesticides have been widely used for a long time for enhancing the food production and improving public health by destroying the insects and pests of crops. Although usage of OCP has been regulated in several countries for agricultural application, they are still being used to kill vectors of human and animal diseases like malaria, dengue, encephalitis, filariasis, etc. This indiscriminate use of OCPs has resulted in their presence and persistence in the environment, leading to their accumulation in the food chain which finally make their way into the human body. Although the majority of the OCP compounds have been restricted or banned in Europe since the 1970s, some OCP compounds including endosulfan remain in legal use or have only recently been banned in few countries.

OCPs are fat-soluble pesticides that may persist and bioaccumulate in the environment, even though their use has been banned or restricted in the past several decades [12]. Contamination by persistent organic pollutants (POPs) is potentially harmful to organisms at higher trophic levels in the food chain. Humans are mainly exposed to these compounds through ingestion since diet is the most significant source of chronic exposure to low doses of these substances [13]. Of the 24 chemicals targeted by the Stockholm Convention, 15 are OCPs, including endosulfan [14]. Today, OCPs have been banned for agricultural or domestic uses in Europe, North America, and some South American countries in agreement with the Stockholm Convention. Nevertheless, some of these compounds are still being applied [15]. Endosulfan (hexachloro-hexahydro-methano-benzodioxathiepinoxide) is an OCP insecticide belonging to class cyclodiene. It is a cyclic sulfurous acid ester with a molecular formula $C_9H_6O_3Cl_6S$ and molecular weight 406.904 [16]. Endosulfan sulfate is the principal metabolite of endosulfan. Other metabolites include endosulfan diol, endosulfan ether, endosulfan hydroxycarboxylic acid, endosulfan hydroxy ether, and endosulfan lactone [17]. Endosulfan sulfate is usually included with the alpha and beta isomers of endosulfan as 'total endosulfan', or 'endosulfan (sum)' in measurements of residues. The endosulfan sulfate is regarded as being equally toxic and of increased persistence in comparison with the parent isomers [18]. Several studies have reported that endosulfan has endocrine-disrupting properties [15, 19, 20]. Because of their persistence in the environment, endosulfan and their metabolites can usually be detected in biological samples [14, 21-23]. In case of endosulfan intoxication, whether accidental or intentional, the ingestion of this compound proves to be a potent neurotoxin for central nervous system and is lethal unless immediate, aggressive treatment is initiated [14]. Human biomonitoring (HBM) is a useful tool for assessing exposure to pesticides and involves the measurement of a biomarker of exposure (usually the pesticide or its metabolites) in human blood (serum and plasma), urine or tissues, thus determining the internal dose of the specific toxicant [24]. There are several problems involved in the analysis of endosulfan and/or their metabolites in biological samples, the most important ones are the low levels found and the complexity of human matrices [23]. Therefore, reliable analytical methods are required for the accurate determination of endosulfan and/or their metabolites in biological specimens [25].

2. ANALYSIS OF ENDOSULFAN AND ITS METABOLITES IN HUMAN SAMPLES

2.1. Sample Preparation

A number of techniques have been reported for the analysis of endosulfan and/or their metabolites in biological samples [26]. In general, all the methods reviewed here share a common scheme, which comprises two key steps: (i) sample pretreatment with the aim of separation, pre-concentration, cleanup and (ii) instrumental quantification [5]. In the sample pretreatment steps, three different techniques comprising liquid-liquid extraction [27-29], solid-phase extraction [30, 31], and microextraction [25, 30] have been described for the isolation of the endosulfan compounds from the biological sample matrices. In this step, simplification, miniaturization, and improvement of the sample extraction and cleanup steps need more attention as they are most time- and labor-extensive parts of the analysis [5,

26, 32, 33]. These methodologies for analysis of endosulfan and its metabolites have been applied to diverse biological samples such as serum, plasma, whole blood, and cord blood/serum, tissues (adipose tissue and placenta, breast milk and urine. In instrumental quantification steps, the gas chromatography (GC) equipped with electron capture detector (ECD), mass spectrometry (MS) and tandem MS (MS/MS) are the most widely reported techniques for the determination of endosulfan.

Table 1. The selected mass-to-charge ratio (m/z) of endosulfan and its metabolites in selected ion monitoring (SIM) mode by GC-MS [27, 66]

Pesticide	SIM mode (m/z)	MS/MS (m/z)	
	Selected ions	Parent ion	Fragment ion
Endosulfan α	195, 239, 241, 277, 279, 339	241	206, 204, 172, 170
Endosulfan β	195, 277, 279	241	206, 204, 172, 170
Endosulfan ether	239, 241, 277	241	239, 206, 204, 172, 170
Endosulfan lactone		321	
Endosulfan sulfate	229, 239, 241, 272, 387	272, 289	
Endosulfan diol	31, 69, 229		

2.1.1. Liquid-Liquid Extraction (LLE)

Liquid-liquid extraction (LLE) is the widely accepted technique for extracting the endosulfan and their metabolites in biological samples. The general procedure consists of shaking biological samples several times with selected organic solvents, followed by centrifugation, ultrasonication, volume reduction, cleanup and subsequent analysis by GC-ECD/MS. In the case of endosulfan compounds, the selection of suitable solvents and sample preparation method is critical to achieving an acceptable recovery from the matrix. Due to their lipophilicity, organic solvents normally can extract endosulfan from biological samples efficiently, but other non-target compounds may interfere and co-extracted with endosulfan. In general, the extraction procedure is followed by a cleanup step to ensure that the co-extractives compounds do not interfere with the analysis of the target compound [34]. For extracting blood samples, ethanol, methanol (MeOH), ethyl acetate (EtAc), hexane, diethyl ether, chloroform, acetone, isopropanol, n-pentane and some mixtures like hexane/MeOH/Isopropanol, acetone/hexane, and hexane/diethyl ether are some solvents that have been broadly used to during LLE. The mixture diethyl ether/hexane and acetone/hexane are the most reported extracting solvent for the blood samples. Table 2 summarizes some of the available techniques for sample preparation and analysis of endosulfan in blood samples. In adipose tissue samples, hexane and a mixture of chloroform/MeOH 1:1 (v/v) have been used to perform liquid extraction. Table 3 shows the sample preparation and analytical parameters employed for the determination of endosulfan in adipose tissue. In placenta tissue samples, solvents such as hexane and mixtures of hexane/acetone have been used to perform LLE. The sample preparation methods and instrumental measurement techniques for the determination of endosulfan in placenta samples have been discussed in Table 4. For extracting breast milk samples, solvents such as hexane [35], MeOH [36], dichloromethane (DCM) [37] and/or mixtures of solvent, such as acetone/hexane [38-41], ethyl ether/hexane [42], acetone/petroleum ether [43], hexane/diethyl ether [36] chloroform-MeOH [44], EtAc/MeOH/acetone [45], DCM/hexane [46, 47], diethyl ether/petroleum ether [48, 49] and cyclohexane/acetone [50] have been more commonly used to perform LLE. The acetone/hexane and DCM/hexane mixtures are the most widely used solvent for extracting the endosulfan in breast milk samples. One of the studies regarding human breast milk samples reported the use of DCM/hexane to perform accelerated LLE [51]. The extraction techniques suitable for breast milk samples are listed in table 5. In some past studies, sonication and/or vortex are also applied to improve extraction efficiency and recoveries. During LLE, it is relatively common to add salts (ex. anhydrous $MgSO_4$) to the solution, in order to improve the separation between the organic and the aqueous phases. In some cases, prior to the LLE, an acid is added to the samples as it allows the determination of free and conjugated pesticides.

Table 2. Extraction and analytical techniques for the determination of endosulfan in serum, plasma, blood and umbilical cord samples

Sample	Sample location	Extraction method	Detection method	Analytes	Unit	LOD	LOQ	Recoveries %	Ref
Serum	Agricultural areas - Almeria (Spain)	LLE Solvents: MeOH, hexane/diethyl ether (1:1, v/v) Reconstituted: hexane/diethyl ether (1:1, v/v) and H$_2$SO$_4$ Solvent: hexane	GC-ECD GC/MS/MS	Endosulfan ether Endosulfan lactone Endosulfan α Endosulfan β Endosulfan sulfate	µg/L	0.12 0.15 0.13 0.15 0.13	0.5 0.5 0.5 0.5 0.5	60–99	[27]
	Spain	Direct immersion SPME at ambient temperature for 45 min Fiber: PDMS	GC-ECD GC/MS	Endosulfan α Endosulfan β Endosulfan ether Endosulfan sulfate	ng/mL	2 3 1 5	NA	70–128	[22]
	Spain	Head-space SPME at 90°C for 30 min. Fiber: Polyacrilate (PA)	GC-ECD GC/MS/MS	Endosulfan α Endosulfan ether	ng/mL	0.5 0.1	NA	NA	[25]
	Castellon province	Head-space SPME at 90°C for 30 min. Fiber: PDMS	GC/MS	Endosulfan α Endosulfan ether Endosulfan sulfate	ng/mL	6 0.5 6	NA	80–119	[58]
	South Spain	LLE - Solvents: MeOH, hexane/diethyl ether (1:1, v/v) Reconstituted: hexane/diethyl ether (1:1, v/v) and H$_2$SO$_4$ Solvent: hexane	GC-ECD GC/MS	Endosulfan α, Endosulfan β, Endosulfan diol, Endosulfan sulfate, Endosulfan lactone, Endosulfan ether	ng/mL	NA	NA	NA	[67]

Table 2. (Continued)

Sample	Sample location	Extraction method	Detection method	Analytes	Unit	LOD	LOQ	Recoveries %	Ref
Serum	Southern Spain	LLE - Solvents: ethyl ether/hexane (1:1, v/v), H_2SO_4 Reconstituted: hexane Cleanup: silica Sep-Pak (Wat 051900) Elution: hexane, hexane:MeOH:isopropanol (45:40:15, v/v/v) Reconstituted: hexane	GC-ECD GC/MS	Endosulfan α, Endosulfan β, Endosulfan diol, Endosulfan sulfate	ng/mL	0.1-3	NA	85-93	[68]
	Four areas in Biscay (Spain)	The solid phase of 500 μL of serum on C18 extraction discsCleanup: liquid-solid adsorption in silica/sulphuric acid columns Reconstituted: cyclohexane	GC/MS	Endosulfan β	NA	NA	NA	60-120	[69]
	Rio Janeiro, Brasil	SPE: PACKs C-18 Conditioned: hexane, MeOH and water Cleanup: Florisil columns Elution: hexane, hexane: ether (6%)	GC-ECD	Endosulfan α, Endosulfan β	ng/mL	0.02	NA	80-98	[70]
	Sonora, Mexico	LLE - Solvents: hexane Reconstituted: hexane	GC-ECD	Endosulfan	μg/L	0.2-0.5	NA	84-101	[71]
	Punjab India	LLE - Solvents: MeOH, hexane: diethyl ether (1:1 v/v) Cleanup: Florisil column Elution: diethyl ether and hexane Reconstituted: hexane: acetone (1:1 v/v)	GC/MS	Endosulfan β	ng/mL	1 ng/g	NA	85.4-95.5	[72]

Sample	Sample location	Extraction method	Detection method	Analytes	Unit	LOD	LOQ	Recoveries %	Ref
Serum	Canada	LLE - Solvents: ethyl ether: hexane solution (1:1, v/v) Cleanup: Florisil column and anhydrous MgSO4	Dual column GC-ECD	Endosulfan α	µg/kg	0.01	NA	NA	[1]
	Mexico	Head-space SPME at 80°C for 50 min Fiber: PDMS	GC/MS	Endosulfan α, Endosulfan β, Endosulfan sulfate	ng/mL	0.22-5.41	NA	68-120	[2]
	Almería, Spain	SPE: C18 Conditioned: MeOH, buffer solution (pH=7) water Elution: hexane: diethyl ether mixture (80:20, v/v). Reconstituted: hexane	GC/MS/MS	Endosulfan α	ng/mL	14	51	93-100	[66]
				Endosulfan β		15	55		
				Endosulfan sulfate		10	34		
				Endosulfan lactone		19	68		
				Endosulfan ether		6	26		
	Castellon province, Spain	SPE: C18 Conditioned: MeOH, MTBE, deionized water Elution: MTBE Reconstituted: hexane	GC/MS/MS	Endosulfan α	ng/mL	0.3	0.9	68-91	[31]
				Endosulfan β		0.5	1.5	84-89	
				Endosulfan sulfate		22	66	63	
				Endosulfan ether		0.1	0.3	80-97	
	Southern Spain	LLE - Solvents: MeOH, hexane/diethyl ether (1:1, v/v) Reconstituted: hexane/diethyl ether (1:1, v/v) and H2SO4 Solvent: hexane	GC-ECD GC/MS/MS	Endosulfan α, Endosulfan β, Endosulfan diol, Endosulfan sulfate, Endosulfan lactone, Endosulfan ether	ng/mL	NA	NA	NA	[67]

Table 2. (Continued)

Sample	Sample location	Extraction method	Detection method	Analytes	Unit	LOD	LOQ	Recoveries %	Ref
Serum	Southern Spain	LLE - Solvents: MeOH, hexane/diethyl ether (1:1, v/v) Reconstituted: hexane/diethyl ether 1:1, v/v) and H_2SO_4 Solvent: hexane	GC/MS/MS	Endosulfan α	µg/L	0.3	NA	76-100	[73]
				Endosulfan β		0.62			
				Endosulfan sulfate		0.2			
				Endosulfan lactone		0.15			
				Endosulfan ether		0.05			
	Coimbra, Portugal	LLE - Solvents: hexane:acetone (9:1, v/v) Cleanup: Florisil column with anhydrous Na_2SO_4 Elution: hexane, hexane: DCM	GC-ECD	Endosulfan sulfate	µg/L	NA	NA	NA	[74]
	Southern Spain	LLE - Solvents: ethyl ether/hexane (1:1 v/v), H_2SO_4 Reconstituted: hexane	GC-ECD	Endosulfan α	ng/mL	NA	0.5	94-100	[75]
				Endosulfan β			2		
				Endosulfan sulfate			0.5		
	Southern Spain	LLE - Solvents: ethyl ether/hexane (1:1, v/v), H_2SO_4 Reconstituted: hexane Cleanup: silica Sep-Pak (Wat 051900) Elution: hexane, hexane:MeOH:isopropanol (45:40:15, v/v/v) Reconstituted: hexane	GC-ECD GC/MS	Endosulfan α, Endosulfan β, Endosulfan diol, Endosulfan sulfate	ng/mL	0.1-3	NA	85-93	[68]
	Cairo, Egypt	SPE: C18 Conditioned: 0.1M HCl solution Wash: 0.1M HCl solution, MeOH:water (1:9 v/v) Elution: MeOH:water (1:1 v/v)	GC/HRMS	Endosulfan α, Endosulfan β	mg/L	NA	NA	NA	[65]

	Sample location	Extraction method	Detection method	Analytes	Unit	LOD	LOQ	Recoveries %	Ref
Plasma	Rio Janeiro, Brasil	LLE - Solvents: MeOH, hexane: diethyl ether (1:1 v/v) Cleanup: Florisil column with anhydrous Na_2SO_4 Elution: hexane, hexane and diethyl ether 6%	GC-ECD	Endosulfan α, Endosulfan β	ng/mL	0.02	NA	NA	[76]
	Karachi, Pakistan	LLE - Solvents: MEOH, hexane and diethyl ether, H_2SO_4 Reconstituted: hexane Cleanup: Florisil column with anhydrous Na_2SO_4 Reconstituted: hexane	GC-ECD	Endosulfan	mg/kg	NA	NA	NA	[23]
	Delhi, India	LLE - Solvents: hexane, acetone (1:1) Cleanup: Florisil column	GC-ECD	Endosulfan α, Endosulfan β	pg/mL	4	NA	NA	[77]
	Malaria Endemic Communities Chiapas, México	LLE - Solvents: absolute ethyl alcohol, ammonium sulfate, and hexane (1:1:3, v/v/v) Cleanup: Florisil column Conditioned: DCM, acetone, and hexane Elution: DCM:hexane (30:70)	GC-ECD	Endosulfan α, Endosulfan β, Endosulfan sulfate	µg/L	0.55-1.14	1.78-3.79	80-102	[78]
	China	SPE: OASIS HLB Wash: DCM, MeOH, deionized water Elution: DCM:hexane (1:9,v/v) Reconstituted: hexane	GC/MS	Endosulfan α Endosulfan sulfate	ng/mL	2.31 0.28	NA	72-119	[79]
	San Luis Potosí, México	LLE - Solvent: hexane Cleanup: florisil column Elution: DCM:hexane (3:7, v/v)	GC-ECD GC/MS	Endosulfan α Endosulfan β	ng/g ng/mL	0.22 0.16	0.74 0.53	NA	[80]

Table 2. (Continued)

Sample	Sample location	Extraction method	Detection method	Analytes	Unit	LOD	LOQ	Recoveries %	Ref
Plasma	Rio Janeiro, Brasil	LLE - Solvents: MeOH, hexane: diethyl ether (1:1 v/v) Cleanup: Florisil column with anhydrous Na$_2$SO$_4$ Elution: hexane, hexane and diethyl ether 6%	GC-ECD	Endosulfan	μg/L	0.02	NA	80-98	[81]
Blood	Ghana	LLE: Solvents: phosphate buffer (pH 7.4), chloroform, hexane Drying: anhydrous Na$_2$SO$_4$ Wash: acetonitrile, DCM, phosphate buffer (pH 6), distilled water and saturated sodium sulfate solution Reconstituted: hexane	GC-ECD	Endosulfan α Endosulfan β Endosulfan sulfate	μg/kg	NA	0.05 0.01 0.01	80–120	[6]
Blood	NA	LLE: Solvents: hexane:acetone (9:1, v/v) Reconstituted: hexane:acetone (9:1, v/v)	GC-ECD	Endosulfan α, Endosulfan β, Endosulfan sulphate, Endosulfan diol	ng/mL	NA	NA	92-96	[82]
Blood	Kerala, India	LLE: Solvents: hexane:acetone (9:1, v/v) Reconstituted: hexane:acetone (9:1, v/v)	GC/MS	Endosulfan α, Endosulfan β, Endosulfan sulphate	pg/mL	NA	NA	98-112	[34]

Sample	Sample location	Extraction method	Detection method	Analytes	Unit	LOD	LOQ	Recoveries %	Ref
Blood	Japan	LLE- Solvents: ethanol/hexane (1:3) and hexane-saturated acetonitrile. Wash: water with sodium sulfuric anhydride Cleanup: florisil column Elution: hexane	GC/MS	The mixture of α and β - Endosulfan	pg/g	NA	NA	NA	[83]
	Delhi, India	LLE - Solvents: hexane, acetone Cleanup: Florisil column	GC-ECD	Endosulfan α	ng/mL	NA	NA	>95	[84]
	three South African Indian Ocean coastal regions	SPE: OASIS HLB Extraction: DCM Reconstituted: hexane Elution through a deactivated silica column with hexane: dichloromethane (9:1) and DCM	GC/MS	Endosulfan α	pg/mL	24	NA	90-117	[30]
				Endosulfan β		2			
	Rural área, San Luis Potosí, Mexico	LLE - Solvents: hexane, Cleanup: Florisil column Elution: DCM:hexane (3:7, v/v) Reconstituted: hexane	GC-ECD	Endosulfan α	ng/mL	0.22	0.74	NA	[80]
				Endosulfan β		0.16	0.53		
	Delhi, India	LLE - Solvents: hexane, acetone Cleanup: Florisil column Elution and Reconstituted: hexane	GC-ECD	Endosulfan α, Endosulfan β	μg/L	4	NA	>95	[85]

Table 2. (Continued)

Sample	Sample location	Extraction method	Detection method	Analytes	Unit	LOD	LOQ	Recoveries %	Ref
Blood	Punjab India	LLE - Solvents: MeOH, hexane: diethyl ether (1:1 v/v) Cleanup: Florisil column and anhydrous Na_2SO_4 Elution: hexane, 6% diethyl ether in hexane, 15% diethyl ether in hexane, 50% diethyl ether in hexane, diethyl ether Reconstituted: hexane: acetone (1:2 v/v)	GC-ECD	Endosulfan β	ng/mL	1 ng/g	NA	65-110	[86]
	Turkey	LLE - Solvents: ethanol, diethyl ether, n-pentane Cleanup 1: silica, alumina with 3% H_2O, and 2 g anhydrous Na_2SO_4 column Elution: hexane and DCM (1:1) Cleanup 2: SPE C18 Elution and Reconstituted: acetonitrile	HRGC/MS	Endosulfan α, Endosulfan β	pg/g (lipid)	NA	NA	50-150	[60]
	Delhi, India	LLE - Solvents: hexane, acetone. Cleanup: Florisil column and anhydrous Na_2SO_4 Elution and Reconstituted: hexane	GC-ECD	Endosulfan	ng/mL	NA	NA	NA	[87]
	Delhi, India	LLE - Solvents: hexane and acetone (1:1) Cleanup: Florisil column	GC-ECD	Endosulfan α, Endosulfan β	pg/mL	4	NA	>95	[88]
	Yucatan, Mexico	SPE: C18 Conditioned: MeOH, deionized water Elution: MTBE Reconstituted: hexane	GC-ECD	Endosulfan α, Endosulfan β, Endosulfan sulphate	µg/mL	NA	NA	NA	[89]

Sample location	Extraction method	Detection method	Analytes	Unit	LOD	LOQ	Recoveries %	Ref
Umbilical cord - serum	Texas, USA	LLE Solvent: hexane Drying: anhydrous Na$_2$SO$_4$ Reconstituted: hexane	GC-ECD GC/MS/MS	Endosulfan β and Endosulfan sulfate	μg/L	0.3	NA	NA	[28]
	Almeria, Spain	LLE - Solvents: MeOH, hexane/diethyl ether (1:1, v/v) Reconstituted: hexane/diethyl ether (1:1, v/v) Cleanup with H$_2$SO$_4$ Reconstituted: hexane	GC-ECD GC/MS	Endosulfan α; Endosulfan β, Endosulfan ether, Endosulfan lactone, Endosulfan diol, Endosulfan sulfate	ng/mL	NA	NA	94-100	[3]
	Japan	LLE- Solvents: ethanol/hexane (1:3) and hexane-saturated acetonitrile. with sodium sulfuric anhydride Wash: water Cleanup: florisil column Elution: hexane	GC/MS	The mixture of α and β - Endosulfan	pg/g	NA	NA	NA	[83]
	Southern Spain	LLE - Solvents: ethyl ether/hexane (1:1, v/v), H$_2$SO$_4$ Reconstituted: hexane Cleanup: silica Sep-Pak (Wat 051900) Elution: hexane, hexane:MeOH:isopropanol (45:40:15, v/v/v) Reconstituted: hexane	GC-ECD GC/MS	Endosulfan α, Endosulfan β	ng/mL	0.1-3	NA	NA	[29]

Table 2. (Continued)

Sample	Sample location	Extraction method	Detection method	Analytes	Unit	LOD	LOQ	Recoveries %	Ref
Umbilical cord - serum	Granada	LLE - Solvents: ethyl ether: hexane (1:1 v/v), hexane with H_2SO_4 Cleanup: silica Sep-Pak Elution: hexane: MeOH: isopropanol (45:40:15; v/v/v) Reconstituted: hexane	GC-ECD GC/MS	Endosulfan α Endosulfan β Endosulfan diol Endosulfan sulfate Endosulfan lactone Endosulfan ether	ng/mL	0.3 0.62 0.05 0.15 0.2 0.2	NA	NA	[90]
	China	SPE: Supelco ENVI-C18	HRGC/MS	Endosulfan α	pg/mL	1	NA	65-110	[64]
	Brittany, France	SPE: Strata-X Extraction: DCM Cleanup: activated florisil column Elution: hexane:DCM (3:1) and DCM Reconstituted: hexane	GC/MS	Endosulfan α	µg/L	0.01-0.05	NA	NA	[91]
	Southern Spain	LLE - Solvents: ethyl ether/hexane (1:1 v/v), H_2SO_4 Reconstituted: hexane	GC-ECD	Endosulfan α Endosulfan β Endosulfan sulfate	ng/mL	NA	0.5 2 0.5	94-100	[75]
	Shanghai, China	SPE: C18 Conditioned: MeOH, MTBE, deionized waterThe sample in water–1-propanol (85:15, v/v) Wash: water–1-propanol (85:15, v/v) Elution: hexane–DCM (1:1, v/v) Reconstituted: hexane Cleanup: florisil-silica gel column, H_2SO_4 Wash: hexane Elution: 10% DCM–hexane, Isooctane	GC-µECD	Endosulfan α, Endosulfan β, Endosulfan sulfate	µg/L	0.06-0.20	NA	63-120	[61]

	Sample location	Extraction method	Detection method	Analytes	Unit	LOD	LOQ	Recoveries %	Ref
Umbilical cord blood	Malaysia	SPE: ISOLUTE C18 preconditioned: MeOH, phosphate buffer (0.04 M, pH 2) Eluent: DCM:ethyl acetate (1:1, v/v) Reconstituted: DCM:ethyl acetate (1:1, v/v)	GC/MS	Endosulfan α, Endosulfan β, Endosulfan sulfate	ng/mL	0.1	0.25	65-120	[92]
	Delhi, India	LLE - Solvents: hexane, acetone Cleanup: Florisil column	GC-ECD	Endosulfan α, Endosulfan β	ng/mL	NA	NA	>95	[84]
	Rural área, San Luis Potosí, Mexico	LLE - Solvents: hexane, Cleanup: Florisil colums Elution: DCM:hexane (3:7, v/v) Reconstituted: hexane	GC-ECD	Endosulfan α	ng/mL	0.22	0.74	NA	[80]
				Endosulfan β		0.16	0.53		
Umbilical cord plasma	Huaihe River Basin	SPE: OASIS HLB Wash: DCM, MeOH, deionized water Elution: DCM:hexane (1:9,v/v) Reconstituted: hexane	GC/MS	Endosulfan α Endosulfan β Endosulfan sulfate	ng/mL	0.08-2.31	0.26-7.69	72-119	[93]
	San Luis Potosí, Mexico	LLE - Solvent: hexane Cleanup: florisil column Elution: DCM:hexane (3:7, v/v)	GC-ECD	Endosulfan α	ng/g	0.22	0.74	NA	[80]
			GC/MS	Endosulfan β	ng/mL	0.16	0.53		

Table 3. Extraction and analytical techniques for determination of endosulfan in adipose tissue samples

Country	Extraction method	Detection method	Analyte	Unit	LOD	LOQ	Recoveries %	Ref
Spain (Granada/Almeria)	LLE - Solvent: hexane Separation: preparative (eluted with hexane and mixture hexane:MeOH:2-isopropanol (40:45:15) (v/v/v) Reconstituted: hexane	GC-ECD GC/MS	Endosulfan ether	ng/mL	0.1	0.5	85-90	[94]
			Endosulfan lactone		2	5		
			Endosulfan diol		0.5	1		
			Endosulfan α		0.5	1		
			Endosulfan β		2	10		
			Endosulfan-sulphate		0.5	5		
Spain (Granada/Almeria)	LLE - Solvent: hexane Separation: preparative HPLC (eluted with hexane and mixture hexane:MeOH:2-isopropanol (40:45:15) (v/v/v) Reconstituted: hexane	GC-ECD GC/MS	Endosulfan ether	ng/mL	0.1	NA	≥85	[67]
			Endosulfan lactone		2			
			Endosulfan diol		0.5			
			Endosulfan-sulphate		0.5			
			Endosulfan α		2			
			Endosulfan β		0.5			
Spain	Drying: anhydrous Na_2SO_4 LLE + vortex Solvent: hexane Cleanup: LC system using the six-way injection valve	GC-ECD GC/MS/MS	Endosulfan α	ng/g	10	NA	60-100	[95]
			Endosulfan ether		5			
			Endosulfan -β		5			
			Endosulfan sulfate		5			
			Endosulfan lactone		5			
Spain (Granada or Almeria)	LLE Solvent: hexane Separation: preparative HPLC (eluted with hexane and mixture hexane:MeOH:2-isopropanol (40:45:15) (v/v/v) Reconstituted: hexane	GC/MS/MS	Endosulfan ether	µg/L	0.05	0.25	76-98	[96]
			Endosulfan lactone		0.15	0.5		
			Endosulfan α		0.3	1		
			Endosulfan β		0.62	2		
			Endosulfan sulfate		0.2	0.75		

Country	Extraction method	Detection method	Analyte	Unit	LOD	LOQ	Recoveries %	Ref
Spain (Almeria/Granada)	LLE Solvent: hexane Separation: preparative HPLC (eluted with hexane and mixture hexane:MeOH:2-isopropanol (40:45:15) (v/v/v) Reconstituted: hexane	GC-ECD GC/MS	Endosulfan α, Endosulfan β, Endosulfan ether, Endosulfan lactone, Endosulfan diol, Endosulfan sulfate	ng/g	NA		94- 100	[3]
Spain (Granada and Almeria)	LLE Solvent: hexane Separation: preparative HPLC (eluted with hexane and mixture hexane:MeOH:2-isopropanol (40:45:15) (v/v/v) Reconstituted: hexane	GC-ECD GC/MS	Endosulfan α Endosulfan β Endosulfan ether Endosulfan lactone Endosulfan diol Endosulfan-sulphate	ng/g	NA		NA	[97]
Spain (Poniente and Granada)	LLE Solvent: hexane Cleanup 1: Alumina Merck column Elution: hexane Cleanup 2: silica Sep-Pak Elution: hexane and hexane: MeOH: isopropanol (45:40:15; v/v/v) Reconstituted: hexane	GC-ECD GC/MS	Endosulfan α Endosulfan sulfate Endosulfan β	ng/mL	NA	0.5 0.5 2	99.79 100 94	[98]
Turkey	LLE Solvents: chloroform/MeOH 1:1 (v/v); Addition of water Separation: Centrifugation	GC-ECD	Endosulfan α Endosulfan β	ng/g lipid	NA	NA	NA	[99]

Table 3. (Continued)

Country	Extraction method	Detection method	Analyte	Unit	LOD	LOQ	Recoveries %	Ref
Portugal	Reconstituted: hexane with H_2SO_4 SPE: Strata C18-E Conditioned and Elution: hexane Dried under the flow of nitrogen Reconstituted: hexane	GC-ECD GC/MS	Endosulfan α Endosulfan β	μg/kg adipose tissue	8.6 5.6	NA	70-79 70-87	[55]

Table 4. Extraction and analytical techniques for the determination of endosulfan in placenta samples

Sample location	Extraction method	Detection method	Analytes	Unit	LOD	LOQ	Recoveries %	Ref
Finland and Denmark	Sample homogenization: Na_2SO_4 and sea sand. LLE Solvents: acetone and hexane (2:1 v/v). Reconstituted: toluene. Cleanup: gel permeation chromatography with toluene eluent in a glass cartridge (packed with alumina B, Na_2SO_4, florisil, silica gel, and Na_2SO_4 from bottom to top). Condensed to about 10 mL toluene	HRGC–HRMS	Endosulfan α Endosulfan β	ng/g lipid	Denmark - 0.01-0.22; Finland 0.01-0.15 NA	NA NA	NA NA	[62]

Sample location	Extraction method	Detection method	Analytes	Unit	LOD	LOQ	Recoveries %	Ref
Spain	LLE solvent: hexane Cleanup: SPE (glass column with alumina Merck 90) Elution: hexane Separation: preparative HPLC Reconstituted: hexane	GC-ECD	Endosulfan ether, Endosulfan lactone, Endosulfan diol, Endosulfan sulfate, Endosulfan α, Endosulfan β	ng/mL	0.1 - 3	NA	84-102	[100]
Finland and Denmark	Sample preparation has been described elsewhere [101]	GC/MS	Endosulfan α	ng/g lipid	NA	NA	NA	[102]
Finland	LLE Solvents: acetone and hexane (2:1 v/v). Reconstituted: toluene. Cleanup: gel permeation chromatography (Bio-Beads S-8 column with toluene eluent) and then in a glass cartridge (packed with alumina B, Na$_2$SO$_4$, florisil, silica gel, and Na$_2$SO$_4$ from bottom to top). Condensed to about 10 mL toluene	GC/MS	Endosulfan α Endosulfan β	ng/g lipid	0.01-0.15 0.01-0.25	NA	52-54	[101]
Spain (Granada and Almeria provinces)	LLE Solvent: hexane Separation: preparative HPLC (eluted with hexane and mixture hexane:MeOH:2-isopropanol (40:45:15) (v/v/v) Reconstituted: hexane	GC-ECD GC/MS	Endosulfan ether, Endosulfan lactone, Endosulfan diol, Endosulfan sulfate, Endosulfan α, Endosulfan β	ng/g placenta	NA		94-100	[3]

Table 4. (Continued)

Sample location	Extraction method	Detection method	Analytes	Unit	LOD	LOQ	Recoveries %	Ref
US	NA	HRGC-ECD	Endosulfan β, Endosulfan sulfate	µg/L	0.3	NA	NA	[28]
China	LLE Solvent: hexane/acetone solvent mixture (1:1 v/v) extracted three times by ultrasonication and vortex mixing. Cleanup: gel permeation chromatography, followed by silica gel column chromatography	GC/MS	Endosulfan α	ng/g lipid	NA	NA	84-96	[103]
Granada /Spain	LLE Solvent: hexane Clean-up: glass column filled with Alumine Solvent: hexane. Separation: preparative HPLC	GC-ECD GC/MS	Endosulfan ether, Endosulfan lactone, Endosulfan diol, Endosulfan sulphate, Endosulfan α, Endosulfan β	ng/mL	NA	0.1-3.0	84-102	[104]

Table 5. Extraction and analytical techniques for measurement of endosulfan in breast milk samples

Sample location	Extraction method	Detection method	Analytes	Unit	LOD	LOQ	Recoveries %	Ref
Egypt	LLE Solvents: acetone:hexane (3:4 v/v), acetone:hexane (1:2 v/v) with H_2SO_4 Reconstituted: hexane	GC–ECD GC/MS	Endosulfan α		NA	NA	>95	[38]
Southern Spain	LLE Solvents: ethyl ether/hexane (1:1 v/v), concentrated sulfuric acid. Cleanup: silica Sep-Pak with hexane and hexane: MeOH: isopropanol (45:40:15; v/v/v). Reconstituted: hexane	GC–ECD GC/MS	Endosulfan α, Endosulfan β, Endosulfan ether, Endosulfan lactone, Endosulfan diol, Endosulfan sulfate		NA	NA	NA	[42]
Ghana	LLE Solvents: acetone/petroleum ether (1:1 v/v), petroleum ether. Cleanup: sulfuric acid, silica gel with petroleum ether. Elution: petroleum ether (HCB, DDE); diethyl ether/petroleum ether (DDT and the rest). Reconstituted: hexane	GC-ECD	Endosulfan α Endosulfan β Endosulfan sulfate	µg/kg	NA	0.01–0.05	78–108	[43]
Indonesia	LLE Solvents: hexane: acetone (2:1 v/v), 2% Na_2SO_4. Cleanup: Florisil cartridges Reconstituted: hexane	GC-ECD	Endosulfan α	mg/kg	0.01	0.01	53–109	[39]

Table 5. (Continued)

Sample location	Extraction method	Detection method	Analytes	Unit	LOD	LOQ	Recoveries %	Ref
India	LLE Solvents: hexane, 10% sodium chloride and 15% DCM Reconstituted: hexane	GC-ECD GC/MS	Endosulfan	mg/kg	0.001		93	[35]
Southern Spain	LLE Solvents: MeOH, hexane:diethyl ether (1:1 v/v). Cleanup: hexane with H_2SO_4 Reconstituted: hexane	GC/MS/MS	Endosulfan ether	μg/L	0.05	0.25	76-98	[36]
			Endosulfan lactone		0.15	0.50		
			Endosulfan α		0.30	1		
			Endosulfan β		0.62	2		
			Endosulfan sulfate		0.20	0.75		
Denmark + Finland	LLE Solvents: acetone:hexane (2:1 v/v). Reconstituted: toluene Cleanup: Gel permeation-chromatography followed by sandwich cartridge	GC/MS	Endosulfan α	NA	NA	NA	NA	[40]
Denmark + Finland	LLE Solvents: acetone:hexane (2:1 v/v). Reconstituted: toluene Cleanup: Gel permeation-chromatography followed by sandwich cartridge	GC/MS	Endosulfan α	ng/g lipid	0.01–0.38	NA	NA	[41]
Turkey	LLE Solvents: chloroform:MeOH mixture (1:1 v/v) with H_2SO_4	GC-ECD	Endosulfan α, Endosulfan β	NA	NA	NA	79–110	[44]

Sample location	Extraction method	Detection method	Analytes	Unit	LOD	LOQ	Recoveries %	Ref
Egypt	LLE Solvents: ethyl acetate:MeOH:acetone (2:4:4 v/v/v) Cleanup: SPE C18-cartridges with ethyl acetate:MeOH:acetone (2:4:4 v/v/v) Wash: acetonitrile Reconstituted: isooctane	GC/MS	Endosulfan α, Endosulfan β	NA	NA	NA	NA	[45]
Turkey	LLE Solvents: DCM:hexane (2:1 v/v). Reconstituted: hexane	GC-ECD	Endosulfan α, Endosulfan β	ng/g lipid	0.004	0.013	89-101	[46]
India	Extraction: activated silica gel with anhydrous Na₂SO₄ and DCM Elution: DCM:acetone (1:2 v/v) Reconstituted: hexane:acetone (1:1 v/v)	GC-ECD and flame thermionic detector	Endosulfan β, Endosulfan sulfate	ng/g	1-2	NA	81-114	[105]
Japan	LLE Solvents: hexane, DCM:hexane (1:1 v/v). Cleanup 1: GPC with a Bio-Beads S-X3 column, Elution: DCM:hexane (1:1 v/v), KOH:ethanol (7:3), hexane, 20% diethylether in hexane Cleanup 2: silica-gel column, Elution: DCM:hexane (12:88 v/v)	GC/MS	Endosulfan α	pg/mL	NA	10	87-99	[47]
Brazil	Extraction:Solid-phase dispersion technique with Celite® Solvents: hexane:acetone (1:1 v/v), hexane:DCM (4:1 v/v) Reconstituted: toluene	GC-ECD	Endosulfan β	μg/mL	0.006	0.013	67-120	[106]

Table 5. (Continued)

Sample location	Extraction method	Detection method	Analytes	Unit	LOD	LOQ	Recoveries %	Ref
China	Extraction: accelerated solvent extraction (ASE) Solvents: DCM:hexane (1:1 v/v). 90 °C, 1,500 psi, 5 min of heat time, 5 min of static time, and 3 static cycles. Reconstituted: cyclohexane:ethyl acetate (1:1 v/v) Cleanup: GPC system, mobile phase: cyclohexane:ethyl acetate (1:1 v/v). Followed by SPE with Florisil cartridges Conditioned and eluted: DCM: hexane (4:1 v/v)	GC/MS	Endosulfan α, Endosulfan β, Endosulfan sulfate Endosulfan lactone Endosulfan ester	ng/kg lipid	1.32–32	NA	60–124	[51]
Jordan	LLE Solvents: 25% ammonia, diethyl ether and petroleum ether (40–60°C) Drying: anhydrous Na_2SO_4 Cleanup: Florisil Reconstituted: hexane	GC-ECD	Endosulfan α, Endosulfan β	mg/L	0.005	NA	79.4	[48]
Mexico	LLE Solvents: sodium chloride (30%, w/ v), DCM Reconstituted: hexane Cleanup: LC-Florisil cartridge Conditioned: hexane Elution: DCM Reconstituted: hexane	GC-ECD	Endosulfan α	ng/mL	10–21	30–64	NA	[37]
			Endosulfan β					
			Endosulfan sulfate					

Sample location	Extraction method	Detection method	Analytes	Unit	LOD	LOQ	Recoveries %	Ref
Tanzania	LLE Solvents: cyclohexane:acetone (3:2 v/v). Cleanup: GPC, filled with Bio-Beads S-X3, 200–400 mesh, and ≥97.5% H_2SO_4	HRGC-LRMS	Endosulfan α, Endosulfan β, Endosulfan sulfate	ng/g	0.023 –0.63	NA	153-196	[50]

2.1.2. Solid-Phase Extraction (SPE)

Solid-phase extraction (SPE) technique is commonly used as an alternative extraction/cleanup method to LLE for the extraction of contaminants in liquid samples [52]. The SPE procedure is further divided in four key steps: conditioning (the functional groups of the sorbent bed are solvated in order to make them interact with the sample), retention (the specific analytes are bound to the bed surface), selective washing (undesired species are removed) and elution (the selective analytes are desorbed and collected) [53].

The selection of an appropriate sorbent depends on the interaction mechanisms between the sorbent and the analytes [15]. To date, typical SPE materials are modified silica with C8, C18, CN and other groups, graphitized carbon black (GCB) and styrene-divinylbenzene copolymers (PS–DVB). SPE with Cl8 cartridges has been used to determine OCPs [54]. For blood samples, SPE with C18 cartridges has been widely reported followed by hydrophilic-lipophilic balance (HLB) and reversed phase SPE (strata X). The solvents such as DCM, MeOH, hexane, diethyl ether, and EtAc and some different mixtures of these were reported for the elution step of SPE in blood samples (Table 2). For adipose tissue samples, only one occasion SPE was applied using strata C18-E [55], with hexane being the solvent applied for the elution step (Table 3).

2.1.3. Solid Phase Microextraction (SPME)

Solid phase microextraction (SPME) technique is a very simple and efficient solventless sample preparation technique developed by Pawliszyn and co-workers in the early 1990s [56]. The advantage of this extraction technique is simplicity or capability of injecting the whole extracted sample [33, 57] and they can be summarized as a significant reduction of sample treatment/manipulation [22]. There are two typical SPME applications, sampling gases headspace (HS) or sampling solutions. The polymer coating acts like a sponge, concentrating the analytes by absorption/adsorption processes.

Kinetics of the SPME extraction process depends on a number of parameters (e.g., film thickness, agitation of the sample and sampling times) [33]. In the case of endosulfan determination in biological samples, HS approach has been preferred in order to avoid interferences from the matrix [22]. Basically, for SPME procedures, a small volume and an appropriate fiber are required. Several commercial fibers with different polarities, such as polydimethylsiloxane (PDMS), polydimethylsiloxane–divinylbenzene (PDMS–DVB), polyacrylate (PA), carbowax–divinylbenzene (CW–DVB), and carboxen–polydimethyl-siloxane (CX–PDMS), among others [15] are available in the market. For endosulfan and their metabolites analysis, HS-SPME with PDMS [2, 22, 58] and PA fibers [25] were reported only for blood samples analysis (Table 2).

2.2. Cleanup Procedures

Even though the analytes of interest are isolated from the bulk matrix, several interfering compounds (fats, proteins, etc.) may also be co-extracted simultaneously with target analytes, which could interfere during instrumental analysis [53, 59]. Additionally, co-extracted compounds, especially lipids, tend to adsorb in GC apparatus mainly in the injector and column, resulting in poor chromatographic performance [13]. Consequently, a purification step is required before the instrumental analysis. For the biological samples, several purification procedures have been performed to eliminate co-extracted interference from extracts including adsorption chromatography, gel permeation chromatography, and solid-phase extraction.

Most of the published methods used anhydrous $MgSO_4$ at one or more steps in order to remove water traces from the extraction solvent system [13]. The most commonly used cleanup procedure is adsorption chromatography applied the SPE technique [60]. In classic processes, the analytes and interfering compounds are adsorbed on a solid sorbent, followed by elution of the target compounds from the substrate and arrest of the interfering substances. In this process, handmade columns are filled with modified silica gel, florisil or alumina. On another hand, the commercial SPE cartridges are also employed. Florisil and silica columns are commonly used, while a mixture of florisil-silica gel [61], florisil-alumina, and silica-alumina [60] has been also reported [62].

2.3. Analytical Techniques

2.3.1. Gas Chromatography (GC)

The potentially toxic nature of pesticide residues and the consequent implementation of strict regulatory guidelines have led the search for fast, sensitive and reliable analytical methods to determine the target compounds at trace levels [34]. Most pesticides are volatile and thermally stable. In the case of OCPs, most of these compounds are non-polar and easily vaporized. Thus, GC is the most common technique for chromatographic separation. Only two detectors coupled with GC have been reported for analysis of endosulfan and their metabolites, i.e., ECD and MS. GC–ECD is the most commonly used method with acceptable detection limits. However, even though the ECD detector is frequently used for quantification, GC/MS detector has gained increasing attention in recent year due to quantification and confirmation. Further, to increase the confidence in the confirmative analysis, a GC coupled with tandem MS is one of the suitable techniques [63]. MS is a very sensitive, universal, yet selective detection technique, and suitable for trace level identification of endosulfan and their metabolites. The tandem MS technique (MS/MS) allows highly specific MS analysis and thereby can improve detection limits by avoiding most of the interferences, especially when analyzing complex matrices such as biological samples [5]. Based on the available literature, the selected ions used for GC/MS and GC/MS/MS are summarized in Table 1. Some examples of the high resolution of GC (HRGC/MS) applications have also been reported for the analysis of these compounds in biological samples [60, 64, 65]. Tables 2, 3, 4 and 5 discuss the analytical parameters used for the blood, adipose tissue, placenta and breast milk, respectively.

3. LEVEL OF ENDOSULFANS IN HUMAN SAMPLES

3.1. Blood, Serum, Plasma and Umbilical Cord Blood Samples

Blood has been the ideal matrix for HBM as it is in contact with all tissues and in equilibrium with organs and tissues. Several studies reported in the literature have revealed some difficulties due to the complexity of the substrate for the determination of residual endosulfan in the blood [34, 66]. Furthermore, most of the analysis is based on the separation and identification of residues in blood serum or in plasma rather than in whole-blood samples. This may be due to the difficulty of eliminating matrix-associated impurities from whole blood samples and due to the separation of plasma from blood [34]. However, blood sampling is an invasive procedure and suffers from ethical and practical constraints, particularly for small children or other susceptible populations. The use of non-invasively collected matrices can be a valuable alternative.

Using cord blood as a non-invasive matrix for biomonitoring offers the advantage that at the same time, the exposure history of both the mother and the early exposure of the newborn infant is studied [107]. A total of 50 original research papers listed in Table 6 showed different studies in several countries about the presence of endosulfan in whole blood, serum, plasma, and cord blood samples. Spain and India were the two countries that published a higher number of studies related endosulfan. México, China, Brazil, Pakistan, USA, Ghana, Malaysia, Japan, Canada, France, Portugal, Egypt, South Africa, and Turkey also showed their contribution in the related field. About 15 research paper conducted in Spain reported concentrations of endosulfan and their metabolites in adult's serum. Furthermore, of four studies reported in agricultural areas region of Spain, only two studies detected endosulfan [25, 27, 58]. In Spain the study that presented higher levels of endosulfan α, β and sulfate was reported in the cord and maternal serum of 72 females [75]. Studies on endosulfan from India mainly focused on the presence of -endosulfan α and β in populations with pathologies [77, 84, 108]. These studies show that endosulfan levels are higher in the populations with pathology compared to the control samples [77, 84, 85, 88, 108]. Among different studies described in table 5, a study from Mexican reported the highest levels for endosulfan in maternal blood (endosulfan α with mean 153 µg/L) and umbilical cord plasma (endosulfan α with mean 401 µg/L) [80]. A study carried out on 18 women suffering from uterine cervix cancer showed endosulfan α with a mean of 43.9 µg/L. This value is higher when compared to the others presented at table 6. The presence of endosulfan compounds was also reported in children from Mexico [2, 71]. The studies from China showed the presence of endosulfan α, endosulfan β and endosulfan sulfate in the range of <LOD-2.62 µg/L, 0.19-0.50 µg/L and 0.11-0.27 respectively [61, 64, 79, 93]. A higher level of endosulfan α (Mean 10 .4 µg/L) [70] was reported in serum samples from Brazil. Serum samples from Egypt infants showed a higher level of endosulfan α and β compared with other studies [65]. No trace of endosulfan compounds was detected in studies from other countries (Ghana, Malaysia, and France) [6, 91, 92].

3.2. Human Tissues

3.2.1. Adipose Tissue

Considering lipophilic behavior of endosulfan it is obvious that this compound may accumulate in body compartments with a higher lipid component such as the adipose tissue [96]. The majority of the available studies related to endosulfan (and/or its metabolites) has been performed on the Spanish population, particularly in regions with intensive agricultural practices such Granada and Almeria provinces located in the south of Spain [3, 67, 94-98]. Except for a single study by Moreno et al. (2004) with no positive samples, all the other studies carried out on Spanish population detected the concentration of endosulfan or its metabolites in the adipose tissue samples (Table 7). Generally, the values, expressed in ng/g of lipid/fat ranged from values below 1 until a maximum value of 944.1 ng/g fat for the α- isomer [3]. Two other studies that were carried out in different populations, one in Turkey [99] and another one in Portugal [55] also found a high level of endosulfan. In the study with the Turkish population, authors tried to assess a correlation between the levels of endosulfan (α and β) and fertility in men [99], but no statistical significance was obtained. For the Portuguese population, the authors found that some organochlorine pesticides (OCPs) were detected in 80 to 95% of the analyzed samples. The endosulfan compounds were not detected in the Portuguese population [55]. Women in reproductive age were the main assessed population in most of the studies. Considering that these compounds can be mobilized during pregnancy, the determination of its presence in other pregnancy-related tissues such as placenta can be fundamental for understanding how children are exposed [67].

3.2.2. Placenta

Concentrations of foreign chemicals in placental tissue have been used to evaluate exposure caused by occupational and environmental sources [62, 102, 112]. Placental tissue is considered a valuable and readily available source of human tissue for biomonitoring, according to research [113]. The comparison between two studies with Finish and Danish mothers (placentas) showed higher values for endosulfan α (the isomer β was not detected) in placentas from Finland. In two studies comparing Finish and Danish mothers (placentas) [62, 102] higher values for endosulfan α (the isomer β was not detected) were found for the Finnish placentas (Table 8). However, the β isomer was detected at lower rates and concentration values, in another study that included only with Finish mothers [114]. Three studies assessed the levels of endosulfan and its metabolites in the Spanish population [3, 100, 104] showed the presence of all the metabolites. The endosulfan diol showed the higher levels. In Spain, a study with 50 newborns with the diagnosis of cryptorchidism and/or hypospadias showed higher levels of endosulfan compounds when compared with controls newborns [100]. In another case-control study [103], in China regarding the risk of neural tube defects, again the cases with pathology achieved higher values. Thus, the presence of endosulfan in placentas can be regarded as a risk factor for the development of malformations in the newborns/child.

3.2.3. Breast Milk

Endosulfan is an environmental persistent pollutant that can pass from mother to child through the placenta during pregnancy or after birth via breastfeeding [45]. Therefore, it is important to monitoring human breast milk. Table 9 summarizes studies dealing with contamination of endosulfan and its metabolites in breast milk samples. The reported concentrations were converted to the same unit (μg/L), allowing a fair comparison. Among different continents, Asia had the highest number of studies related to endosulfan contamination in breast milk [39, 44, 46, 47, 51, 105].

A study performed in India in 2011 comparing breast milk from primiparous and multiparous mothers, showed the maximum values found for endosulfan β and endosulfan sulfate, with a concentration values ranging between ND - 145.2 μg/L, and ND - 22.41 μg/L respectively. The primiparous mothers aged between 20 and 38 years old presented the higher values [105]. Another study carried out in India presented the highest value for endosulfan with a concentration of 763 μg/L [35]. Other studies were made in breast milk from mothers between 18 and 39 years old regarding the evaluation of endosulfan α [39, 46, 47]. The maximum value of 0.7 μg/L was found in Indonesian mothers [39].

Regarding endosulfan lactone the values ranged from ND to 1.01 μg/L, the maximum value was found to be present in breast milk from women between 18 and 35 years old from China [51]. One study was performed comparing primiparous and multiparous mothers from Antalya, Turkey in a region with agricultural production during the four seasons [44]. In this case, the primiparous mothers presented the higher values, 0.384 μg/L and 4.8 μg/L, for endosulfan α and β respectively [44]. No values were detected for endosulfan in the studies performed in Jordan [48, 49]. A study from Spain detected endosulfan in breast milk. The higher value (3.93 μg/L) was found for endosulfan ether in a study analyzing three stadiums of breast milk (colostrum, transition and mature).

The maximum value was present in mature breast milk from women between 17 and 35 years old [42]. A study comparing breast milk from mothers living in Demark or Finland showed that Finland mothers presented higher value for endosulfan α (0.965 µg/L) [41]. Still, about these countries, another study was made comparing the breast milk from mothers giving birth healthy and cryptorchid boys [40]. Despite the incidence of cryptorchidism has been linked to endocrine-disrupting chemicals, endosulfan α concentration (0.838 µg/L) was higher in mothers giving bird healthy boys [40].

Africa continent presented studies from Tanzania and Ghana where no values of endosulfan compounds were detected [43, 50]. In the case of Egypt, two studies were performed. In 1993, the maximum value for endosulfan α (61.80 µg/L) was found in breast milk from women between 20 and 30 years old [38]. The breast milk from the OCPs intoxicated mothers revealed the higher values found in this sample (51.8 µg/L and 29.05µg/L for α- and β-endosulfan) [45]. A study from Brazil showed higher concentration (18.91 µg/L) for endosulfan β in breast milk from mothers between 18-49 years old [106]. In all continents where the studies were made the authors found the presence of endosulfan or/and its metabolites, showing the wide distribution of this OCP in the environment worldwide.

3.2.4. Urine Samples

In HBM studies urine is not the matrix of choice for persistent chemicals [115]. Although some compounds can/could be detected in participants' urine in a few studies, these levels typically occurred around the limit of detection. Nevertheless, few studies have reported the presence of endosulfan and its metabolites in urine samples in farmers from rural areas of Spain, with intensive agricultural practices [116, 117]. The results showed that the endosulfan concentrations were higher in urine samples of the workers immediately post pesticide application than the worker with several hours of endosulfan application (Table 10).

Table 6. Levels of endosulfan and their metabolites in serum, plasma, blood and umbilical cord blood samples

Country	Sample description		Endosulfan and metabolites levels (µg/L)						Ref
			Endosulfan α	Endosulfan β	Endosulfan ether	Endosulfan lactone	Endosulfan sulphate	Endosulfan diol	
Spain	serum	Agricultural areas donors	ND	ND	ND	ND	ND	NA	[27]
		Agricultural workers	ND	NA	1	NA	ND	NA	[58]
		NA	86	ND	ND	NA	35	NA	[22]
		Agricultural workers (N=9)	Mean 4.94-14.54	Mean 2.77-6.86	NA	Mean 0.18	NA	NA	[66]
		Agricultural workers	ND	NA	ND	NA	NA	NA	[25]
		Postmenopausal women (N=200)	Mean 1.72; Max 7.27	Mean 7.66; Max 35.92	Mean 1.66; Max 12.77	Mean 0.76 Max 3.03	Mean 2.01; Max 8.47	Mean 12.81; Max 180.37	[67]
		Umbilical cord, average age 29 years old (17-43 years) (N=200)	Median 1.56	Median 2	Median 0.81	Median 2.07	Median 1.20	Median 9.62	[3]
		Males 18-23 years (N=220)	Mean 2.10; Max 19.39	Mean 1.31; Max 6.85	NA	NA	Mean 2.17; Max 53.32	Mean 15.39; Max 76.86	[68]
		Umbilical cord (N= 318)	Mean 2.44; Max 15.15	Mean 2.83; Max 7.63	NA	NA	NA	NA	[29]
		2–75 years (N= 162)	NA	ND	NA	NA	NA	NA	[109]
		Donors (N=10)	ND	ND	ND	ND	ND	NA	[31]
		Agricultural areas -women 35–70 years old	0.014	ND	ND	ND	ND	NA	[73]
		Umbilical cord (N=72)	Mean 3.18	Mean 66.05	NA	NA	Mean 31.83	NA	[75]
		Maternal donors (N=72)	Mean 2.1	Mean 76.38	NA	NA	Mean 34.17	NA	

Country	Sample description			Endosulfan and metabolites levels (µg/L)						Ref
				Endosulfan α	Endosulfan β	Endosulfan ether	Endosulfan lactone	Endosulfan sulphate	Endosulfan diol	
India		blood	Males (N=200)	Mean 2.1	Mean 1.31	NA	NA	NA	Mean 15.39	[68]
			Umbilical cord (N= 320)	Mean 2.44	Mean 2.83	NA	Mean 10.63	Mean 2.59	Mean 0.92	[90]
			Male/female donors >10 years (N= 106)	ND	ND	NA	NA	NA	NA	[34]
			Maternal donors: full-term delivery (FTD); preterm delivery (PTD)	Mean FTD 2.10; Mean PTD 3.10	Mean FTD 1.22; Mean PTD 1.43	NA	NA	NA	NA	[84]
			Umbilical cord FTD and PTD	Mean FTD 2.10; Mean PTD 3.10	Mean FTD 2.10; Mean PTD 3.10	NA	NA	NA	NA	
		plasma	Healthy (H) and chronic kidney disease (CKD) donors 30-50 years (N=150)	Mean H 0.001; Mean CKD 0.004	Mean H 0.002; Mean CKD 0.004	NA	NA	NA	NA	[77]
		blood	Healthy (H) (N=75) and Alzheimer's disease (AD) donors (N=70)	Mean H 0.001; Mean AD 0.003	Mean control 0.001; Mean AD 0.002	NA	NA	NA	NA	[108]
			Healthy and CKD donors (N_{total}=270)	Mean H 1.33; Mean CKD 3.85	Mean H 1.71; Mean CKD 3.27	NA	NA	NA	NA	[85]
			Donors (N=111)	NA	Mean 34.90	NA	NA	NA	NA	[86]

Table 6. (Continued)

Country	Sample description			Endosulfan and metabolites levels (µg/L)						Ref
				Endosulfan α	Endosulfan β	Endosulfan ether	Endosulfan lactone	Endosulfan sulphate	Endosulfan diol	
			Mothers-neonate dyads and neonates (N=70)	Median 0.0-1.9	NA	NA	NA	NA	NA	[87]
			CKD patients (N=200)	0.007-0.002	0.008-0.002	NA	NA	NA	NA	[88]
		serum	Donors (N=50)	NA	Mean 4.08	NA	NA	NA	NA	[72]
Mexico		serum	Children 6-12 years (N=165)	Mean 0.25	NA	NA	NA	NA	NA	[71]
		plasma	Maternal donors (N=50)	Mean 153	Mean 90	NA	NA	NA	NA	[80]
			Umbilical cord (N=50)	Mean 401	Mean 265	NA	NA	NA	NA	
		serum	Children (6-12 years) (N=247)	Median 1.5	Median 2.4	NA	NA	Median 1.3	NA	[2]
		blood	Uterine cervix cancer donors (N=18)	Mean 43.9	Mean 1.6	NA	NA	Mean 0.2	NA	[89]
China		serum	Mother-infant pairs maternal and umbilical cord (N=81)	Mean <LOD Max 0.001	NA	NA	NA	NA	NA	[64]
			The control group (N=121) Mothers who delivered neural tube defects (NTD) infants (N=117)	ND	ND	ND	ND	ND	ND	[110]
		plasma	Umbilical cord (N=972)	Mean 1.73	Mean 0.50	NA	NA	Mean 0.27	NA	[93]
			Pregnant women (N=120)	Mean 2.62	NA	NA	NA	Mean 0.26	NA	[79]
		serum	Umbilical cord (N=1438)	Mean 1.51	Mean 0.19	NA	NA	Mean 0.11	NA	[61]

Country	Sample description		Endosulfan and metabolites levels (µg/L)						Ref
			Endosulfan α	Endosulfan β	Endosulfan ether	Endosulfan lactone	Endosulfan sulphate	Endosulfan diol	
Brazil	plasma	Pregnant women (N=64)	Arithmetic mean 0.108	NA	NA	NA	NA	NA	[81]
	serum	Children (N=193)	Mean 10.4	Mean 3.10	NA	NA	NA	NA	[70]
		Males (N=303)	Median 0.22	Median 0.23	NA	NA	NA	NA	[76]
		Women (N=305)	Median 0.22	Median 0.24	NA	NA	NA	NA	
Pakistan		Farm workers (N=56)	0.41-1.96	NA	NA	NA	NA	NA	[111]
-		Cancer patients (N=83)	0.17-0.21	NA	NA	NA	NA	NA	[23]
USA	blood	Umbilical cord from farmworkers (N=9)	NA	ND	NA	NA	0.9	NA	[28]
Ghana		Farming community: men (N=10), women (N=10)	ND	ND	NA	NA	ND	NA	[6]
Malaysia		Umbilical cord (N=180)	ND	ND	NA	NA	ND	NA	[92]
Japan		Maternal donors (N=32)	Mean 2.90		NA	NA	NA	NA	[83]
	serum	Umbilical cord (N=32)	ND	ND	NA	NA	NA	NA	
Canada		Males (N=9), females (N=11) mean ages 45 years	Range 0.11	NA	NA	NA	NA	NA	[1]
France		Umbilical cord (N=282)	<LOD	NA	NA	NA	NA	NA	[91]
Portugal		University students (N=160)	NA	NA	NA	NA	Mean 42.6	NA	[74]
Egypt		Infants (N=360)	Mean 100	Mean 126	NA	NA	NA	NA	[65]
South Africa	plasma	Maternal donors (three Indian Ocean Coastal zones) (N=241)	Median 0.05-0.07	Median 0.01-0.07	NA	NA	NA	NA	[30]
Turkey	blood	Newborns boys (N=37)	Mean 29.35 pg/g	Mean 1.33 pg/g	NA	NA	NA	NA	[60]

NA – not analyzed/not available, ND – not detected.

Table 7. Levels of endosulfan and their metabolites in adipose tissue around the world

Country	Sample description	Endosulfan and metabolites levels (µg/L)							Unit	Ref
		Endosulfan α	Endosulfan β	Endosulfan ether	Endosulfan lactone	Endosulfan sulfate	Endosulfan diol			
Spain	Postmenopausal women, mean age- 53 years, undergoing surgery: for malignancies (N= 86); bile vesicle disease (N=71); abdominal or inguinal hernia (N= 26); varices (N= 6); or other diseases (N= 11).	Mean 6.02	Mean 73.36	Mean 1.04	Mean 2.02	Mean 12.17	Mean 9.23		ng/g lipid	[67]
	Women with different diseases undergoing surgery, mean age- 53 years, (N=400)	Median 2.54	Median 20.33	Median 0.94	Median 2.02	Median 15.34	Median 5.15		ng/g lipid	[94]
	Women with different diseases undergoing surgery, mean age- 53 (range 35-65) (N=18)	ND	ND	5-12	ND	15-27	ND		ng/g	[95]
	Women aged 18-35 years	ND	ND	ND	ND	ND	NA		mg/kg	[96]
	Women, mean age- 44 years (range 33–57 years) (N=149)	Median 0.5	Median 2.5	Median 0.25	Median 0.5	Median 1.25	Median 1.25		ng/g fat	[3]

Country	Sample description	Endosulfan and metabolites levels (μg/L)						Unit	Ref
		Endosulfan α	Endosulfan β	Endosulfan ether	Endosulfan lactone	Endosulfan sulfate	Endosulfan diol		
	Women (range age between 33–75 years) (N=458)	Mean 1.82	Mean 7.82	Mean 1.79	Mean 1.48	Mean 9.34	Mean 4.79	ng/g lipid	[97]
	Women scheduled for birth by cesarean section (range age from 18 to 35 years) (N=72)	3.03-334.18	7.23-64.47	NA	NA	12.01-153.45	NA	ng/g fat	[98]
Portugal	Patients undergoing bariatric surgery	NA	NA	NA	NA	NA	NA	ng/g	[55]
Turkey	Fertile men (ranged from 21 to 46 years old; mean 34.5 ± 6.4) (N=21) Infertile men (ranged from 25 to 45 years old; mean age 32.7 ± 7.0) (N=25)	Infertile: N.D -0.07; Fertile: N.D - 0.009	Infertile: N.D -0.03; Fertile: N.D -0.99	NA	NA	NA	NA	ng/g lipid	[99]

NA – not analyzed/not available, ND – not detected

Table 8. Levels of endosulfan and their metabolites in placenta samples reported worldwide

Country	Sample description	Endosulfan and metabolites levels in placenta							Unit	Ref
		Endosulfan α	Endosulfan β	Endosulfan ether	Endosulfan lactone	Endosulfan sulfate	Endosulfan diol		
Denmark Finland	Placentas from a joint perspective, a longitudinal birth cohort study	Danish 0.27-6; Finnish 0.26-8.78; Mean Danish - 2.28; Mean Finland 2.52	ND	NA	NA	NA	NA	ng/g lipid	[62]
	Danish mothers (N=43) Finish mothers (N=43)	Mean Danish - 2.28; Mean Finland - 2.52	ND	NA	NA	NA	NA	ng/g lipid	[102]
Finland	Finish mothers	0.3-8.8	0.1-5.6	NA	NA	NA	NA	ng/g lipid	[114]
Spain	Mother-child cohort (N= 702)	cases 0.5-28.3 control 0.5-26.7	cases 2.0-46.2 control 2.0-10.7	cases 0.1-1.01 control 0.1-1.0	cases 0.1-55.3 control 0.1-74.7	cases 0.5-21.9 control 0.5-44.6	cases 0.5-13.8 control 0.5-45.6	ng/g lipid	[100]
	Women, mean age 29 years (range, 17–43 years) (N=200)	Median 0.94	Median 1.00	Median 0.2	Median 2.64	Median 1.61	Median 4.35	ng/g	[3]
	Mother-son pairs (N=220)	GM 0.73	GM 1.37	GM 0.23	GM 1.14	GM 0.93	GM 2.1	ng/g	[104]
US	Pregnant farmworkers (N= 9)	NA	ND	NA	NA	ND	NA	µg/L	[112]
China	Mothers cases (N= 79) Control (N=47)	Case: median 0.079 control: median - 0.054	NA	NA	NA	NA	NA	ng/g lipid	[103]

NA – not analyzed/not available, ND – not detected, GM – Geometric mean.

Table 9. Levels of endosulfan and their metabolites in breast milk samples

Country	Sample description	Endosulfan and metabolites levels (µg/L)						Ref
		Endosulfan α	Endosulfan β	Endosulfan ether	Endosulfan lactone	Endosulfan sulphate	Endosulfan diol	
Egypt	Mothers 20-30 years (N=60)	0.00 - 61.80	NA	NA	NA	NA	NA	[38]
Spain	Colostrum - Mothers 17-35 years	QL	2.3	2.05	L.C.	2.9	QL	[42]
	Transition milk - Mothers 17-35 years	QL - 0.4	3.2	1.85	QL.	3.4	QL	
	Mature milk - Mothers 17-35 years	0.165	2.84	3.93	QL	QL - 1.23	QL	
Ghana	Mothers 23-42 years (N=20)	ND	ND	NA	NA	ND	NA	[43]
Indonesia	Mothers 18-33 years (N=10)	≤0.35 - 0.7	NA	NA	NA	NA	NA	[39]
India	Mothers 19-45 years (N=12)	763						[35]
Spain	Mothers 18-35 years (N=5)	ND	ND	ND	ND	ND	NA	[36]
Denmark + Finland	Mothers giving birth to cryptorchid boys 30-31 years (N=62)	0.07 - 0.66	NA	NA	NA	NA	NA	[40]
	Mothers giving birth to healthy boys 29-31 years (N=68)	0.04 - 0.84	NA	NA	NA	NA	NA	
Denmark	Mothers 23.6-38.8 years (N=65)	0.06 - 0.51	NA	NA	NA	NA	NA	[41]

Table 9. (Continued)

Country	Sample description	Endosulfan and metabolites levels (µg/L)							Ref
		Endosulfan α	Endosulfan β	Endosulfan ether	Endosulfan lactone	Endosulfan sulphate	Endosulfan diol		
Finland	Mothers 21.4-39.7 years (N=65)	0.05 - 0.97	NA	NA	NA	NA	NA		
Turkey	Primiparous mothers 19-31 years (N=50)	N.D - 0.38	N.D - 4.8	NA	NA	NA	NA		[44]
	Multiparous mothers 23-38 years (N=50)	N.D - 0.33	N.D - 1.32	NA	NA	NA	NA		
Egypt	Control mothers 20-40 years (N=180)	8.75	0.88	NA	NA	NA	NA		[45]
	CP-intoxicated mothers 20-40 years (N=180)	51.8	29.05	NA	NA	NA	NA		
Turkey	Mothers 27.8 ± 7.60 years (N=101)	ND - 0.01	NA	NA	NA	NA	NA		[46]
India	Primiparous mothers 20-38 years (N=34)	NA	ND - 145.2	NA	NA	N.D - 22.41	NA		[105]
	Multiparous mothers 20-38 years (N=19)	NA	ND	NA	NA	ND	NA		
Japan	Mothers 26-39 years (N=9)	0.01 - 0.07	NA	NA	NA	NA	NA		[47]
Brazil	Mothers 18-49 years (N=62)		16.74 - 18.91	NA	NA	NA	NA		[106]
China	Mothers 18-35 years (N=142)	ND - 1.38	N.D - 0.42	0.003 - 0.64	ND - 1.01	ND - 4.1			[51]
Jordan	Mothers (N=100)	ND	ND	NA	NA	NA	NA		[48]

Country	Sample description	Endosulfan and metabolites levels (µg/L)							Ref
		Endosulfan α	Endosulfan β	Endosulfan ether	Endosulfan lactone	Endosulfan sulphate	Endosulfan diol		
	Mothers 20-45 years (N=100)	ND	ND	NA	NA	NA	NA		[49]
Mexico	Mothers (N=24)	ND – 2.00	ND – 5.50	NA	NA	NA	NA		[37]
Tanzania	Primiparous mothers (N=95)	below LOD	below LOD	NA	NA	below LOD	NA		[50]

NA – not analyzed/not available, ND – not detected, QL – quantifiable limit.

Table 10. Levels of endosulfan and their metabolites in urine samples around the world

Country	Sample description	Endosulfan and metabolites levels in urine samples							Ref
		Endosulfan α	Endosulfan β	Endosulfan ether	Endosulfan lactone	Endosulfan sulfate	Endosulfan diol	Unit	
Canada	Males 44.5 ± 14.4 years (N=9) Females 45.6 ± 10.3 years(N=10) Donors with various clinical conditions (N=10) Controls (N=10)	0.12-0.18	NA	NA	NA	NA	NA	µg/kg	[118]
Spain	Farmworkers 20-55 years old	last application 1 day ago							[117]
		787-894	801-896	ND -125	ND-222	ND	NA	pg/mL	
		last application 7 days ago							
		n.d-123	n.d-169	ND	ND-428	ND-516	NA	pg/mL	
Spain	Farmworker 34 years (N=1)	1710-4289	491-1079	ND	ND	ND	NA	pg/mL	[116]

NA – not analyzed/not available, ND – not detected.

CONCLUSION

The determination of OCPs such as endosulfan and its metabolites is extremely important to protect the environment and human health since they are a lipophilic compound with high resistance to degradation and long half-lives in humans. Consequently, reliable methods with sufficiently low detection limits are urgently required to support monitoring and regulatory enforcement.

One analytical challenge in this field is to present consistent results, following official guidelines, as fast as possible and considering method characteristics such as recovery, accuracy, sensitivity, and specificity. As is demonstrated in this chapter, the endosulfan and its metabolic products are widespread and can be found in blood, umbilical cord blood, breast milk, and other tissues such as placenta and adipose tissue. Concerning the analytical evaluation of the described studies, most methods involved LLE, SPE, and SPME, with very low LOD and adequate recoveries. In conclusion, the present chapter brings together a set of studies that show the importance of the biomonitoring studies that can be considered as an established approach for the assessment of occupational and environmental exposure in humans of toxic compounds. Therefore, endosulfan and their metabolites have proven to be toxic compounds to which humans are frequently exposed.

ACKNOWLEDGMENTS

The work was supported by UID/QUI/50006/2019 with funding from FCT/MCTES through national funds. Virgínia Cruz Fernandes, Maria Luz Maia are grateful to Fundação para a Ciência e a Tecnologia (FCT) for the Postdoc grant (SFRH/BPD/109153/2015) and PhD grant (SFRH/BD/128817/2017) respectively.

REFERENCES

[1] Genuis SJ, Lane K, Birkholz D. Human Elimination of Organochlorine Pesticides: Blood, Urine, and Sweat Study. *Biomed Res Int.* 2016:10.

[2] Flores-Ramirez R, Perez-Vazquez FJ, Rodriguez-Aguilar M, Medellin-Garibay SE, Van Brussel E, Cubillas-Tejeda AC, et al. Biomonitoring of persistent organic pollutants (POPs) in child populations living near contaminated sites in Mexico. *Sci Total Environ.* 2017;579:1120-6.

[3] Cerrillo I, Granada A, López-Espinosa M-J, Olmos B, Jiménez M, Caño A, et al. Endosulfan and its metabolites infertile women, placenta, cord blood, and human milk. *Environ Res.* 2005;98(2):233-9.

[4] ER L, WJ H. *Handbook of Pesticide Toxicology.* San Diego, CA: Academic Press; 1991.

[5] Margariti MG, Tsakalof AK, Tsatsakis AM. Analytical methods of biological monitoring for exposure to pesticides: recent update. *Ther Drug Monit.* 2007;29(2):150-63.

[6] Ntow WJ. Organochlorine pesticides in water, sediment, crops, and human fluids in a farming community in Ghana. *Arch Environ Contam Toxicol.* 2001;40(4):557-63.

[7] Correia-Sá L, Fernandes VC, Carvalho M, Calhau C, Domingues VF, Delerue-Matos C. Optimization of QuEChERS method for the analysis of organochlorine pesticides in soils with diverse organic matter. *J Sep Sci.* 2012;35(12):1521-30.

[8] Fernandes VC, Domingues VF, Mateus N, Delerue-Matos C. Multiresidue pesticides analysis in soils using modified QuEChERS with disposable pipette extraction and dispersive solid-phase extraction. *J Sep Sci.* 2012;36:376-82.

[9] Fernandes VC, Domingues VF, Mateus N, Delerue-Matos C. Organochlorine Pesticide Residues in Strawberries from Integrated Pest Management and Organic Farming. *J Agric Food Chem.* 2011;59(14):7582-91.

[10] Fernandes VC, Vera JL, Domingues VF, Silva LM, Mateus N, Delerue-Matos C. Mass spectrometry parameters optimization for the 46 multiclass pesticides determination in strawberries with gas chromatography ion-trap tandem mass spectrometry. *J Am Soc Mass Spectrom*. 2012;23(12):2187-97.

[11] Correia-Sá L, Fernandes V, Calhau C, Domingues V, Delerue-Matos C. Optimization of QuEChERS Procedure Coupled to GC-ECD for Organochlorine Pesticide Determination in Carrot Samples. *Food Analytical Methods*. 2012:1-11.

[12] Coakley J, Bridgen P, Bates MN, Douwes J, t Mannetje A. Chlorinated persistent organic pollutants in the serum of New Zealand adults, 2011–2013. *Sci Total Environ*. 2018;615:624-31.

[13] LeDoux M. Analytical methods applied to the determination of pesticide residues in foods of animal origin. A review of the past two decades. *J Chromatogr A*. 2011;1218(8):1021-36.

[14] Menezes RG, Qadir TF, Moin A, Fatima H, Hussain SA, Madadin M, et al. Endosulfan poisoning: An overview. *J Forensic Leg Med*. 2017;51:27-33.

[15] Martins JG, Amaya Chávez A, Waliszewski SM, Colín Cruz A, García Fabila MM. Extraction and clean-up methods for organochlorine pesticides determination in milk. *Chemosphere*. 2013;92(3):233 *Endosulfan* -46.

[16] *PubChem*. 2018 [Available from https://pubchem.ncbi.nlm.nih.gov/compound/ endosulfan.

[17] German Federal Environment Agency (GFEA-U). Endosulfan Draft Dossier Prepared in Support of a Proposal of Endosulfan to be Considered as a Candidate for Inclusion in the CLRTAP Protocol on Persistent Organic Pollutants. In: German Federal Environment Agency B, editor. Berlin2007.

[18] *Endosulfan Readers Guide 16*, EPA-HQ-OPP-2002-0262-0057 (2007).

[19] Pirard C, Compere S, Firquet K, Charlier C. The current environmental levels of endocrine disruptors (mercury, cadmium, organochlorine pesticides, and PCBs) in a Belgian adult population and their

predictors of exposure. *Int J Hyg Environ Health.* 2018;221(2):211-22.

[20] Weiss JM, Bauer O, Blüthgen A, Ludwig AK, Vollersen E, Kaisi M, et al. Distribution of persistent organochlorine contaminants in infertile patients from Tanzania and Germany. *J Assist Reprod Genet.* 2006;23(9-10):393-9.

[21] Sinha SN, Patel TS, Desai NM, Mansuri MM, Dewan A, Saiyed HN. Gas chromatography-mass spectroscopy (GC-MS) study of Endosulfan in biological samples. *Asian J Chem.* 2004;16(3-4):1685-90.

[22] Lopez FJ, Pitarch E, Egea S, Beltran J, Hernandez F. Gas chromatographic determination of organochlorine and organophosphorus pesticides in human fluids using solid phase microextraction. *Anal Chim Acta.* 2001;433(2):217-26.

[23] Attaullah M, Yousuf MJ, Shaukat S, Anjum SI, Ansari MJ, Buneri ID, et al. Serum organochlorine pesticides residues and risk of cancer: A case-control study. *Saudi Journal of Biological Sciences.* 2017.

[24] Ken Sexton LLNaJLP. *Human Biomonitoring of Environmental Chemicals Centers for Disease Control and Prevention (CDC) the USA.* 2004.

[25] Hernandez F, Pitarch E, Beltran J, Lopez FJ. Headspace solid-phase microextraction in combination with gas chromatography and tandem mass spectrometry for the determination of organochlorine and organophosphorus pesticides in whole human blood. *Journal of Chromatography B-Analytical Technologies in the Biomedical and Life Sciences.* 2002;769(1):65-77.

[26] Aprea C, Colosio C, Mammone T, Minoia C, Maroni M. Biological monitoring of pesticide exposure: a review of analytical methods. *J Chromatogr B.* 2002;769(2):191-219.

[27] Vidal JLM, Frias MM, Frenich AG, Olea-Serrano F, Olea N. Trace determination of alpha- and beta-endosulfan and three metabolites in human serum by gas chromatography electron capture detection and gas chromatography-tandem mass spectrometry. *Rapid Commun Mass Spectrom.* 2000;14(11):939-46.

[28] Cooper SP, Burau K, Sweeney A, Robison T, Smith MA, Symanski E, et al. Prenatal exposure to pesticides: A feasibility study among migrant and seasonal farmworkers. *Am J Ind Med*. 2001;40(5):578-85.

[29] Mariscal-Arcas M, Lopez-Martinez C, Granada A, Olea N, Lorenzo-Tovar ML, Olea-Serrano F. Organochlorine pesticides in umbilical cord blood serum of women from Southern Spain and adherence to the Mediterranean diet. *Food Chem Toxicol*. 2010;48(5):1311-5.

[30] Channa KR, Rollin HB, Wilson KS, Nost TH, Odland JO, Naik I, et al. Regional variation in pesticide concentrations in plasma of delivering women residing in rural Indian Ocean coastal regions of South Africa. *J Environ Monit*. 2012;14(11):2952-60.

[31] Pitarch E, Serrano R, Lopez FJ, Hernandez F. Rapid multiresidue determination of organochlorine and organophosphorus compounds in human serum by solid-phase extraction and gas chromatography coupled to tandem mass spectrometry. *Anal Bioanal Chem*. 2003;376(2):189-97.

[32] Barr DB, Needham LL. Analytical methods for biological monitoring of exposure to pesticides: a review. *J Chromatogr B*. 2002;778(1):5-29.

[33] Vas G, Vekey K. Solid-phase microextraction: a powerful sample preparation tool prior to mass spectrometric analysis. *J Mass Spectrom*. 2004;39(3):233-54.

[34] Ramesh A, Ravi PE. Determination of residues of endosulfan in human blood by a negative ion chemical ionization gas chromatographic/mass spectrometric method: impact of long-term aerial spray exposure. *Pest Manage Sci*. 2003;59(3):252-8.

[35] Sanghi R, Pillai MKK, Jayalekshmi TR, Nair A. Organochlorine and organophosphorus pesticide residues in breast milk from Bhopal, Madhya Pradesh, India. *Hum Exp Toxicol*. 2003;22:73-6.

[36] Moreno Frías M, Jiménez Torres M, Garrido Frenich A, Martínez Vidal JL, Olea-Serrano F, Olea N. Determination of organochlorine compounds in human biological samples by GC-MS/MS. *Biomed Chromatogr*. 2004;18:102-11.

[37] Polanco Rodríguez ÁG, Inmaculada Riba López M, Angel DelValls Casillas T, León JAA, Anjan Kumar Prusty B, Álvarez Cervera FJ. Levels of persistent organic pollutants in breast milk of Maya women in Yucatan, Mexico. *Environ Monit Assess.* 2017;189.

[38] Saleh M, Kamel A, Ragab A, El-Baroty G, El-Sebae AK. Regional distribution of organochlorine insecticide residues in human milk from Egypt. *Journal of Environmental Science and Health - Part B Pesticides, Food Contaminants, and Agricultural Wastes.* 1996;31:241-55.

[39] Burke ER, Holden AJ, Shaw IC. A method to determine residue levels of persistent organochlorine pesticides in human milk from Indonesian women. *Chemosphere.* 2003;50:529-35.

[40] Damgaard IN, Skakkebæk NE, Toppari J, Virtanen HE, Shen H, Schramm KW, et al. Persistent pesticides in human breast milk and cryptorchidism. *Environ Health Perspect.* 2006;114:1133-8.

[41] Shen H, Main KM, Andersson AM, Damgaard IN, Virtanen HE, Skakkebaek NE, et al. Concentrations of persistent organochlorine compounds in human milk and placenta are higher in Denmark than in Finland. *Hum Reprod.* 2008;23:201-10.

[42] Campoy C, Jiménez M, Olea-Serrano MF, Moreno Frias M, Cañabate F, Olea N, et al. Analysis of organochlorine pesticides in human milk: Preliminary results. *Early Hum Dev.* 2001;65:183-90.

[43] Ntow WJ. Organochlorine pesticides in water, sediment, crops, and human fluids in a farming community in Ghana. *Arch Environ Contam Toxicol.* 2001;40:557-63.

[44] Çok I, Yelken Ç, Durmaz E, Üner M, Sever B, Satlr F. Polychlorinated biphenyl and organochlorine pesticide levels in human breast Milk from the Mediterranean city Antalya, Turkey. *Bull Environ Contam Toxicol.* 2011;86:423-7.

[45] Schaal MF, Abdelraouf SM, Mohamed WA, Hassanein FS. Correlation between maternal milk and infant serum levels of chlorinated pesticides (CP) and the impact of elevated CP on bleeding tendency and immune status in some infants in Egypt. *J Immunotoxicol.* 2012;9:15-24.

[46] Canbay HS, Ogüt S, Yilmazer M, Unsal RS. Pesticide residues analysis in human milk samples in Isparta Region (Turkey). *Asian J Chem.* 2013;25:3931-6.

[47] Fujii Y, Nishimura E, Kato Y, Harada KH, Koizumi A, Haraguchi K. Dietary exposure to phenolic and methoxylated organohalogen contaminants in relation to their concentrations in breast milk and serum in Japan. *Environ Int.* 2014;63:19-25.

[48] Al Antary TM, Alawi MA, Estityah H, Haddad N. Organochlorine Pesticides Residues in Human Breast Milk from the Middle Governorates in Jordan in 2013/2014. *Bull Environ Contam Toxicol.* 2017;99:89-92.

[49] Al Antary TM, Alawi MA, Othman MA, Haddad N. Levels of organochlorine pesticides residues in human breast milk from the northern districts in Jordan in 2014/2015. *Fresenius Environ Bull.* 2017;26:4711-5.

[50] Müller MHB, Polder A, Brynildsrud OB, Karimi M, Lie E, Manyilizu WB, et al. Organochlorine pesticides (OCPs) and polychlorinated biphenyls (PCBs) in human breast milk and associated health risks to nursing infants in Northern Tanzania. *Environ Res.* 2017;154:425-34.

[51] Lu D, Wang D, Ni R, Lin Y, Feng C, Xu Q, et al. Organochlorine pesticides and their metabolites in human breast milk from Shanghai, China. *Environmental Science and Pollution Research.* 2015;22:9293-306.

[52] Lambropoulou DA, Albanis TA. Methods of sample preparation for determination of pesticide residues in food matrices by chromatography–mass spectrometry-based techniques: a review. *Anal Bioanal Chem.* 2007;389(6):1663-83.

[53] Żwir-Ferenc A, Biziuk M. An Analysis of Pesticides and Polychlorinated Biphenyls in Biological Samples and Foods. *Crit Rev Anal Chem.* 2004;34(2):95-103.

[54] Masqué N, Marcé RM, Borrull F. New polymeric and other types of sorbents for solid-phase extraction of polar organic micropollutants from environmental water. *TrAC, Trends Anal Chem.* 1998;17(6):384-94.

[55] Fernandes VC, Pestana D, Monteiro R, Faria G, Meireles M, Correia-Sá L, et al. Optimization and validation of organochlorine compounds in adipose tissue by SPE-gas chromatography. *Biomedical Chromatography.* 2012;26(12):1494-501.

[56] Pawliszyn J. New directions in sample preparation for analysis of organic compounds. *TrAC, Trends Anal Chem.* 1995;14(3):113-22.

[57] Mirnaghi FS, Pawliszyn J. Reusable solid-phase microextraction coating for direct immersion whole-blood analysis and extracted blood spot sampling coupled with liquid chromatography-tandem mass spectrometry and direct analysis in real time-tandem mass spectrometry. *Anal Chem.* 2012;84(19):8301-9.

[58] Beltran J, Pitarch E, Egea S, Lopez FJ, Hernandez F. Gas chromatographic determination of selected pesticides in human serum by head-space solid-phase microextraction. *Chromatographia.* 2001;54(11-12):757-63.

[59] Gilbert-López B, García-Reyes JF, Molina-Díaz A. Sample treatment and determination of pesticide residues in fatty vegetable matrices: A review. *Talanta.* 2009;79(2):109-28.

[60] Ulutas OK, Cok I, Darendeliler F, Aydin B, Coban A, Henkelmann B, et al. Blood concentrations and risk assessment of persistent organochlorine compounds in newborn boys in Turkey. A pilot study. *Environmental Science and Pollution Research.* 2015;22(24):19896-904.

[61] Cao L-L, Yan C-H, Yu X-D, Tian Y, Zhao L, Liu J-X, et al. Relationship between serum concentrations of polychlorinated biphenyls and organochlorine pesticides and dietary habits of pregnant women in Shanghai. *Sci Total Environ.* 2011;409(16):2997-3002.

[62] Shen H, Main KM, Andersson A-M, Damgaard IN, Virtanen HE, Skakkebaek NE, et al. Concentrations of persistent organochlorine compounds in human milk and placenta are higher in Denmark than in Finland. *Human Reproduction.* 2008;23(1):201-10.

[63] Chung SWC, Chen BLS. Determination of organochlorine pesticide residues in fatty foods: A critical review on the analytical methods and their testing capabilities. *J Chromatogr A.* 2011;1218(33):5555-67.

[64] Guo H, Jin YL, Cheng YB, Leaderer B, Lin SB, Holford TR, et al. Prenatal exposure to organochlorine pesticides and infant birth weight in China. *Chemosphere.* 2014;110:1-7.

[65] Schaalan MF, Abdelraouf SM, Mohamed WA, Hassanein FS. Correlation between maternal milk and infant serum levels of chlorinated pesticides (CP) and the impact of elevated CP on bleeding tendency and immune status in some infants in Egypt. *J Immunotoxicol.* 2012;9(1):15-24.

[66] Arrebola FJ, Vidal JLM, Fernandez-Gutierrez A. Analysis of endosulfan and its metabolites in human serum using gas chromatography-tandem mass spectrometry. *J Chromatogr Sci.* 2001;39(5):177-82.

[67] Botella B, Crespo J, Rivas A, Cerrillo I, Olea-Serrano MF, Olea N. Exposure of women to organochlorine pesticides in Southern Spain. *Environ Res.* 2004;96(1):34-40.

[68] Carreno J, Rivas A, Granada A, Lopez-Espinosa MJ, Mariscal M, Olea N, et al. Exposure of young men to organochlorine pesticides in Southern Spain. *Environ Res.* 2007;103(1):55-61.

[69] Zubero MB, Aurrekoetxea JJ, Murcia M, Ibarluzea JM, Goni F, Jimenez C, et al. Time Trends in Serum Organochlorine Pesticides and Polychlorinated Biphenyls in the General Population of Biscay, Spain. *Arch Environ Contam Toxicol.* 2015;68(3):476-88.

[70] Freire C, Koifman RJ, Sarcinelli P, Rosa AC, Clapauch R, Koifman S. Long term exposure to organochlorine pesticides and thyroid function in children from Cidade dos Meninos, Rio de Janeiro, Brazil. *Environ Res.* 2012;117:68-74.

[71] Meza-Montenegro MM, Valenzuela-Quintanar AI, Balderas-Cortes JJ, Yanez-Estrada L, Gutierrez-Coronado ML, Cuevas-Robles A, et al. Exposure Assessment of Organochlorine Pesticides, Arsenic, and Lead in Children From the Major Agricultural Areas in Sonora, Mexico. *Arch Environ Contam Toxicol.* 2013;64(3):519-27.

[72] Bedi JS, Gill JPS, Kaur P, Sharma A, Aulakh RS. Evaluation of pesticide residues in human blood samples from Punjab (India). *Vet World.* 2015;8(1):66-71.

[73] Frias MM, Torres MJ, Frenich AG, Vidal JLM, Olea-Serrano F, Olea N. Determination of organochlorine compounds in human biological samples by GC-MS/MS. *Biomed Chromatogr.* 2004;18(2):102-11.

[74] Lino CM, da Silveira MI. Evaluation of organochlorine pesticides in serum from students in Coimbra, Portugal: 1997-2001. *Environ Res.* 2006;102(3):339-51.

[75] Torres MJ, Folgoso CC, Reche FC, Velasco AR, Garcia IC, Arcas MM, et al. Organochlorine pesticides in serum and adipose tissue of pregnant women in Southern Spain giving birth by cesarean section. *Sci Total Environ.* 2006;372(1):32-8.

[76] Freire C, Koifman RJ, Sarcinelli PN, Simões Rosa AC, Clapauch R, Koifman S. Long-term exposure to organochlorine pesticides and thyroid status in adults in a heavily contaminated area in Brazil. *Environ Res.* 2013;127:7-15.

[77] Siddharth M, Datta SK, Bansal S, Mustafa M, Banerjee BD, Kalra OP, et al. Study on organochlorine pesticide levels in chronic kidney disease patients: Association with estimated glomerular filtration rate and oxidative stress. *J Biochem Mol Toxicol.* 2012;26(6):241-7.

[78] Ruiz-Suarez LE, Castro-Chan RA, Rivero-Perez NE, Trejo-Acevedo A, Guillen-Navarro GK, Geissen V, et al. Levels of Organochlorine Pesticides in Blood Plasma from Residents of Malaria-Endemic Communities in Chiapas, Mexico. *Int J Env Res Public Health.* 2014;11(10):10444-60.

[79] Luo D, Pu YB, Tian HY, Wu WX, Sun X, Zhou TT, et al. Association of in utero exposure to organochlorine pesticides with thyroid hormone levels in cord blood of newborns. *Environ Pollut.* 2017;231:78-86.

[80] Alvarado-Hernandez DL, Montero-Montoya R, Serrano-Garcia L, Arellano-Aguilar O, Jasso-Pineda Y, Yanez-Estrada L. Assessment of exposure to organochlorine pesticides and levels of DNA damage in mother-infant pairs of an agrarian community. *Environ Mol Mutag.* 2013;54(2):99-111.

[81] Sarcinelli PN, Pereira ACS, Mesquita SA, Oliveira-Silva JJ, Meyer A, Menezes MAC, et al. Dietary and reproductive determinants of plasma

organochlorine levels in pregnant women in Rio de Janeiro. *Environ Res.* 2003;91(3):143-50.

[82] Ramesh A, Ravi PE. A rapid and sensitive analytical method for the quantification of residues of endosulfan in blood. *J Environ Monit.* 2002;4(2):190-3.

[83] Fukata H, Omori M, Osada H, Todaka E, Mori C. Necessity to Measure PCBs and Organochlorine Pesticide Concentrations in Human Umbilical Cords for Fetal Exposure Assessment. *Environ Health Perspect.* 2005;113(3):297-303.

[84] Pathak R, Suke SG, Ahmed T, Ahmed RS, Tripathi AK, Guleria K, et al. Organochlorine pesticide residue levels and oxidative stress in preterm delivery cases. *Hum Exp Toxicol.* 2010;29(5):351-8.

[85] Siddarth M, Datta SK, Mustafa MD, Ahmed RS, Banerjee BD, Kalra OP, et al. Increased level of organochlorine pesticides in chronic kidney disease patients of unknown etiology: Role of GSTM1/GSTT1 polymorphism. *Chemosphere.* 2014;96:174-9.

[86] Sharma A, Gill JPS, Bedi JS. Monitoring of Pesticide Residues in Human Blood from Punjab, India. *Bull Environ Contam Toxicol.* 2015;94(5):640-6.

[87] Kalra S, Dewan P, Batra P, Sharma T, Tyagi V, Banerjee BD. Organochlorine pesticide exposure in mothers and neural tube defects in offsprings. *Reprod Toxicol.* 2016;66:56-60.

[88] Ghosh R, Siddarth M, Singh N, Tyagi V, Kare PK, Banerjee BD, et al. Organochlorine pesticide level in patients with chronic kidney disease of unknown etiology and its association with renal function. *Environ Health Prevent Med.* 2017;22(1):8.

[89] Rodriguez AGP, Lopez MIR, Casillas TAD, Leon JAA, Mahjoub O, Prusty AK. Monitoring of organochlorine pesticides in blood of women with uterine cervix cancer. *Environ Pollut.* 2017;220:853-62.

[90] Monteagudo C, Mariscal-Arcas M, Heras-Gonzalez L, Ibanez-Peinado D, Rivas A, Olea-Serrano F. Effects of maternal diet and environmental exposure to organochlorine pesticides on newborn weight in Southern Spain. *Chemosphere.* 2016;156:135-42.

[91] Warembourg C, Debost-Legrand A, Bonvallot N, Massart C, Garlantezec R, Monfort C, et al. Exposure of pregnant women to persistent organic pollutants and cord sex hormone levels. *Hum Reprod.* 2016;31(1):190-8.

[92] Tan BLL, Mohd MA. Analysis of selected pesticides and alkylphenols in human cord blood by gas chromatograph-mass spectrometer. *Talanta.* 2003;61(3):385-91.

[93] Luo D, Pu YB, Tian HY, Cheng J, Zhou TT, Tao Y, et al. Concentrations of organochlorine pesticides in umbilical cord blood and related lifestyle and dietary intake factors among pregnant women of the Huaihe River Basin in China. *Environ Int.* 2016;92-93:276-83.

[94] Rivas A, Fernandez MF, Cerrillo I, Ibarluzea J, Olea-Serrano MF, Pedraza V, et al. Human exposure to endocrine disrupters: Standardisation of a marker of estrogenic exposure in adipose tissue. *APMIS.* 2001;109(3):185-97.

[95] Hernandez F, Pitarch E, Serrano R, Gaspar JV, Olea N. Multiresidue determination of endosulfan and metabolic derivatives in human adipose tissue using automated liquid chromatographic cleanup and gas chromatographic analysis. *Journal of analytical toxicology.* 2002;26(2):94-103.

[96] Moreno FM, Jiménez TM, Garrido FA, Luis MVJ, Fátima OS, Nicolás O. Determination of organochlorine compounds in human biological samples by GC-MS/MS. *Biomedical Chromatography.* 2004;18(2):102-11.

[97] Cerrillo I, Olea-Serrano MF, Ibarluzea J, Exposito J, Torne P, Laguna J, et al. Environmental and lifestyle factors for organochlorine exposure among women living in Southern Spain. *Chemosphere.* 2006;62(11):1917-24.

[98] Jimenez Torres M, Campoy Folgoso C, Cañabate Reche F, Rivas Velasco A, Cerrillo Garcia I, Mariscal Arcas M, et al. Organochlorine pesticides in serum and adipose tissue of pregnant women in Southern Spain giving birth by cesarean section. *Sci Total Environ.* 2006;372(1):32-8.

[99] Çok I, Durmaz TC, Durmaz E, Satıroglu MH, Kabukcu C. Determination of organochlorine pesticide and polychlorinated biphenyl levels in adipose tissue of infertile men. *Environmental Monitoring and Assessment.* 2010;162(1):301-9.

[100] Fernandez MF, Olmos B, Granada A, López-Espinosa MJ, Molina-Molina J-M, Fernandez JM, et al. Human exposure to endocrine-disrupting chemicals and prenatal risk factors for cryptorchidism and hypospadias: a nested case-control study. *Environmental health perspectives.* 2007;115 Suppl 1(Suppl 1):8-14.

[101] Shen H, Main KM, Kaleva M, Virtanen H, Haavisto AM, Skakkebaek NE, et al. Prenatal organochlorine pesticides in placentas from Finland: Exposure of male infants born during 1997-2001. *Placenta.* 2005;26(6):512-4.

[102] Shen H, Main KM, Virtanen HE, Damggard IN, Haavisto A-M, Kaleva M, et al. From mother to child: Investigation of prenatal and postnatal exposure to persistent bioaccumulating toxicants using breast milk and placenta biomonitoring. *Chemosphere.* 2007;67(9):S256-S62.

[103] Ren A, Qiu X, Jin L, Ma J, Li Z, Zhang L, et al. Association of selected persistent organic pollutants in the placenta with the risk of neural tube defects. *Proceedings of the National Academy of Sciences.* 2011;108(31):12770-5.

[104] Freire C, Lopez-Espinosa M-J, Fernández M, Molina-Molina J-M, Prada R, Olea N. Prenatal exposure to organochlorine pesticides and TSH status in newborns from Southern Spain. *Science of The Total Environment.* 2011;409(18):3281-7.

[105] Bedi JS, Gill JPS, Aulakh RS, Kaur P, Sharma A, Pooni PA. Pesticide residues in human breast milk: Risk assessment for infants from Punjab, India. *Sci Total Environ.* 2013;463-464:720-6.

[106] Palma DCA, Lourencetti C, Uecker ME, Mello PRB, Pignati WA, Dores EFGC. Simultaneous determination of different classes of pesticides in breast milk by solid-phase dispersion and GC/ECD. *J Brazil Chem Soc.* 2014;25:1419-30.

[107] Smolders R, Schramm KW, Nickmilder M, Schoeters G. Applicability of non-invasively collected matrices for human biomonitoring. *Environ Health.* 2009;8:8.

[108] Singh NK, Chhillar N, Banerjee BD, Bala K, Basu M, Mustafa M. Organochlorine pesticide levels and risk of Alzheimer's disease in north Indian population. *Hum Exp Toxicol.* 2013;32(1):24-30.

[109] Zubero MB, Aurrekoetxea JJ, Ibarluzea JM, Goni F, Lopez R, Etxeandia A, et al. Organochlorine pesticides in the general adult population of Biscay (Spain). *Gac Sanit.* 2010;24(4):274-81.

[110] Wang B, Yi DQ, Jin L, Li ZW, Liu JF, Zhang YL, et al. Organochlorine pesticide levels inmaternal serumand risk of neural tube defects in offspring in Shanxi Province, China: A case-control study. *Sci Total Environ.* 2014;490:1037-43.

[111] Saeed MF, Shaheen M, Ahmad I, Zakir A, Nadeem M, Chishti AA, et al. Pesticide exposure in the local community of Vehari District in Pakistan: An assessment of knowledge and residues in human blood. *Sci Total Environ.* 2017;587-588:137-44.

[112] Cooper SP, Burau K, Sweeney A, Robison T, Smith MA, Symanski E, et al. Prenatal exposure to pesticides: a feasibility study among migrant and seasonal farmworkers. *Am J Ind Med.* 2001;40(5):578-85.

[113] Myllynen P, Pasanen M, Pelkonen O. Human placenta: a human organ for developmental toxicology research and biomonitoring. *Placenta.* 2005;26(5):361-71.

[114] Shen H, Main KM, Kaleva M, Virtanen H, Haavisto AM, Skakkebaek NE, et al. Prenatal organochlorine pesticides in placentas from Finland: Exposure of male infants born during 1997–2001. *Placenta.* 2005;26(6):512-4.

[115] Angerer J, Ewers U, Wilhelm M. Human biomonitoring: state of the art. *International Journal of Hygiene and Environmental Health.* 2007;210(3-4):201-28.

[116] Arrebola FJ, Martínez Vidal JL, Fernández-Gutiérrez A. Excretion study of endosulfan in urine of a pest control operator. *Toxicology Letters.* 1999;107(1):15-20.

[117] Martınez Vidal JL, Arrebola FJ, Fernández-Gutiérrez A, Rams MA. Determination of endosulfan and its metabolites in human urine using gas chromatography-tandem mass spectrometry. *Journal of Chromatography B: Biomedical Sciences and Applications.* 1998;719(1):71-8.

[118] Genuis SJ, Lane K, Birkholz D. Human Elimination of Organochlorine Pesticides: Blood, Urine, and Sweat Study. *BioMed Research International.* 2016;2016:10.

In: Endosulfan　　　　　　　　　　ISBN: 978-1-53615-910-3
Editors: I. C. Yadav and N. L. Devi　© 2019 Nova Science Publishers, Inc.

Chapter 5

RESIDUE LEVELS, HEALTH IMPACTS, AND ECO-TOXICITY OF ENDOSULFAN IN TROPICAL ENVIRONMENT

M. D. M. D. W. M. M. K. Yatawara
Department of Zoology and Environmental Management
University of Kelaniya Kelaniya, Sri Lanka

ABSTRACT

Endosulfan, although used for the betterment of human being has ultimately resulted in being dangerous to the environment and humans. The toxic residues present in the environment harm the development and normal functioning of the hormone-dependent processes in flora, and fauna and has led to acute and chronic human poisoning. Except for few countries in tropical belt, the majority has banned the production, import and usage of endosulfan at present. Although extensive studies have been carried out in different geographical areas of the temperate region to assess the residue levels, fate, behavior, and ecotoxicity of endosulfan and its transformation products, limited studies have been contributed to describe its behavior, residue levels and ecotoxicity in tropics. The present chapter attempts to extract almost all possible studies on the residue levels, health impacts and ecotoxicity of endosulfan from different compartments of the environment in tropics to one spot. The study further extracts laboratory investigations in order to fill the gaps of ecotoxicity in the tropical environment

Keywords: endosulfan, ecotoxicity, health impacts, tropical environment

1. INTRODUCTION

Endosulfan (1,2,3,4,7,7-hexachlorobicyclo-2,2,1-heptene-2,3-bis-hydroxy methane-5,6 sulfite) is a broad spectrum chlorinated contact insecticide and acaricide which was first introduced in 1954 by Farbwerke Hoechst, Germany [1]. Technical grade endosulfan is commercially available as a mixture of two diastereoisomers, known as α-endosulfan and β-endosulfan in a ratio of 70:30 (Figure 1). The technical grade product contains at least 94% endosulfan according to the United Nations for Food and Agriculture (FAO Specification 89/TC/S), which stipulated that the α-isomer content is between 64 and 67% and the β-isomer is between 29 to 32% [3]. Endosulfan has become an important agrochemical and pest control agent resulting in its global use to control a range of insect pests on a wide variety of vegetables, fruits, cereal grains, and cotton, as well as ornamental shrubs, trees, vines, and ornamentals for the use in commercial agricultural settings. Endosulfan formulations are emulsifiable concentrate, wettable powder, smoke tablets and ultra-low volume liquid formulations [4] and they are available in common trade names of Thiodan®, Endox®, Thiomul®, Beosit®, Endocell®, Malix®, Thionex®, Insecto®, and Tiovel® [5].

Despite its widespread application in many countries due to low production cost and efficacy for many pests on economically important crops, more than sixty countries have prohibited its use in the decade of 1990 due to its persistence and high toxicity in the environment. The isomers of endosulfan are semi-volatile, with high vapor pressures (Table 1), making them susceptible to volatilization to the atmosphere with subsequent atmospheric transport and deposition [6]. The extensive use of endosulfan in a wide variety of crops has contributed to their significant release into the environment. Monitoring programs have revealed that endosulfan is a ubiquitous environmental contaminant occurring in many environmental compartments with the abundance of reported data on the order of α->β->-sulphate (primary metabolite of endosulfan). The endosulfan has also been found in remote locations and therefore has a tendency to undergo long-range transport. Continuous monitoring has revealed higher concentrations in the Arctic region.

Based on laboratory studies, field studies, modeling, field monitoring, and published papers, the US EPA concluded that endosulfan is a highly persistent chemical that can remain in the environment for a very long period, especially in the acidic environment [7]. Endosulfan is particularly neurotoxic for both insects and mammals, including humans. A number of studies demonstrate the high toxicity of endosulfan and formulations of endosulfan to aquatic organisms, including invertebrates [7]. Endosulfan has been reported as endocrine disruptor [8]. It has also been reported to cause damage to the cell membrane of Red Blood Cells at a concentration of 0.001 g/mL [9]. Moreover, in humans, it causes damage to the circulatory, respiratory, excretory system and even developing fetus [10]. It lowers the viability of human T-cell leukemia cell line and also lowers its growth in a dose and time-dependent manner [11]. Endosulfan both isomers have been reported as a genotoxic agent to human HepG2 cells [12]. One of the criteria for designation of a chemical as a POP has been suggested by UNEP, 2001 to have octanol-water partitioning coefficient (log K_{OW}) > 5 [4]. Log K_{OW} of both isomers of endosulfan and endosulfan sulfate has been found below this value but is quite close to it (Table 1), which make endosulfan likely to be bioaccumulated. Endosulfan has even higher octanol-air partitioning coefficients (Log K_{OA}) meaning that bioaccumulation is greater in terrestrial animals than aquatic life [13]. Extensive studies have been carried out in different geographical areas especially in temperate regions to assess the ecotoxicity of endosulfan and its transformation products.

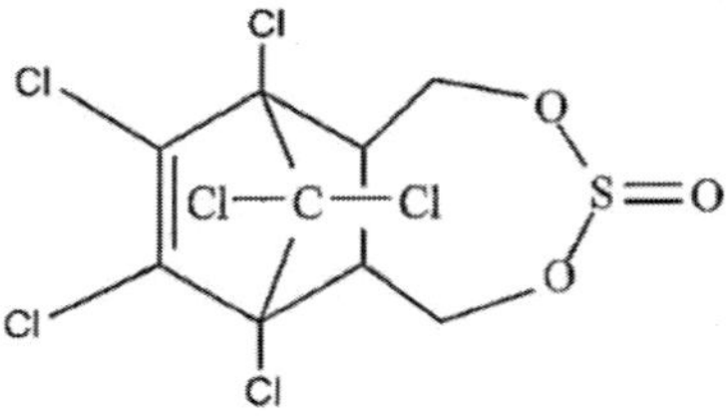

Endosulfan

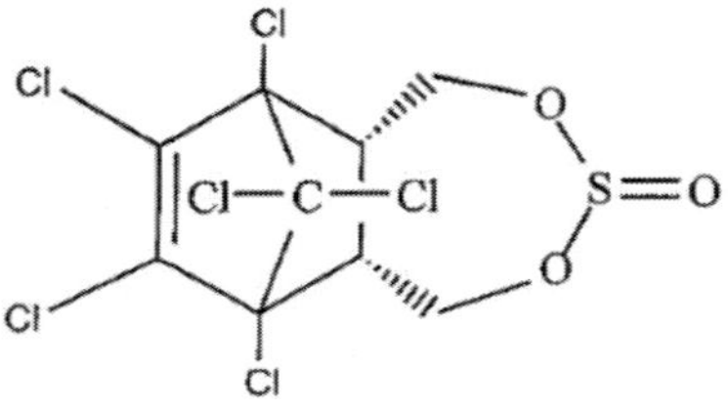

α-Endosulfan 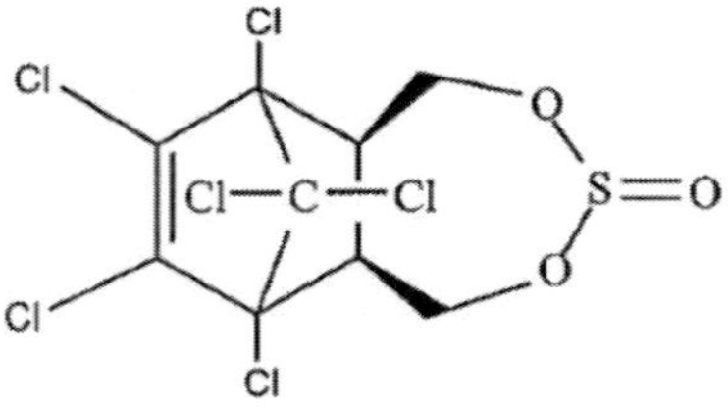   β-Endosulfan

Figure 1. Endosulfan and its stereoisomers, as published by Kumar and Philip, [2].

Table 1. Physicochemical properties of endosulfan isomers and endosulfan sulphate

Parameter	α-endosulfan	β-endosulfan	Endosulfan sulphate
Molecular weight (g/mol)	406.9	406.9	422.9
Vapor pressure (Vp) mPa at 25°C	1.9[a]	0.9[a]	0.037[b]
Water solubility (Sw) mg/L at 25°C	0.32[a]	0.33[a]	0.117[b]
Henry's law constant (H) Pa m^3/mol	0.70[c]	0.045[c]	≈0.015[d]
octanol-water partition coefficient (Log K_{OW})	4.74[e]	4.79[e]	3.77[e]
octanol-air partition coefficient (Log K_{OA})	10.29[d]	10.29[d]	5.8[d]
Sorption coefficient (Log K_{OC})	3.6[f]	4.3[f]	NA

[a]Rosendhal et al., [14]; [b]Ozer, [15]; [c]Shen and Wania, [16]; [d]Weber, et al. [5]; [e]IPEN, [17]; [f]Peterson and Batley, [18].

Although many countries have prohibited its use, endosulfan is still used in China, India, Brazil, and several other countries due to the absence of alternative chemicals that can match its low cost and broad-spectrum efficacy for various insect pests [19-21]. In addition, endosulfan is used for integrated pest management programmes in some countries. Nevertheless, the negative impacts of endosulfan, its stereoisomers and its metabolites (the sulfate, lactone, ether, hydroxy ether, and diol derivatives) on the environment and human health cannot be easily ignored. As indicated, α-isomer is more volatile [22] and thought to be more toxic. Once released to the environment, the parent compound is degraded by abiotic and biotic processes. In the degradation process, the β-isomer is easily converted into α- endosulfan, but not vice versa [23]. In addition, some enzymatic reactions in the presence of certain microbes transform α-endosulfan to the more toxic endosulfan sulfate [24]. On the contrary, hydrolysis of endosulfan in some bacteria (*Pseudomonas aeruginosa*, *Burkholderia cepaeia*) yields the less toxic metabolite endosulfan diol [25]. The endosulfan diol can then be converted to endosulfan ether [26] or endosulfan hydroxy ether [27] and then endosulfan lactone [28]. Hydrolysis of endosulfan lactone yields endosulfan hydroxycarboxylate [29] (Figure 2).

Owing to the intense heat and high humidity in the tropics, endosulfan as the parent compound and its transformation products dissipate at faster rates than in temperate regions. Increase in temperature is known to increase pesticide loss from soils through volatilization, chemical degradation, and bacterial decomposition. But, continuous or illegal applications of endosulfan even in tropical environments may result in significant toxicity to both environment and humans. In addition, repeated application increases pest resistance, while its effects on other species can facilitate the pest's resurgence [31]. Thus, the study of how endosulfan and its transformation products interact with organisms in the environment is much important in future management aspects of endosulfan. Although limited studies have been contributed to confirm the ecotoxicity in different geographical areas in tropics, this chapter attempts to compile all possible scientific investigations on the residue levels, health impacts and ecotoxicity of endosulfan from different compartments in tropical region to one spot. In addition, the study further extracts laboratory ecotoxicity investigations in order to fill the gaps of ecotoxicity of endosulfan in the tropical environment.

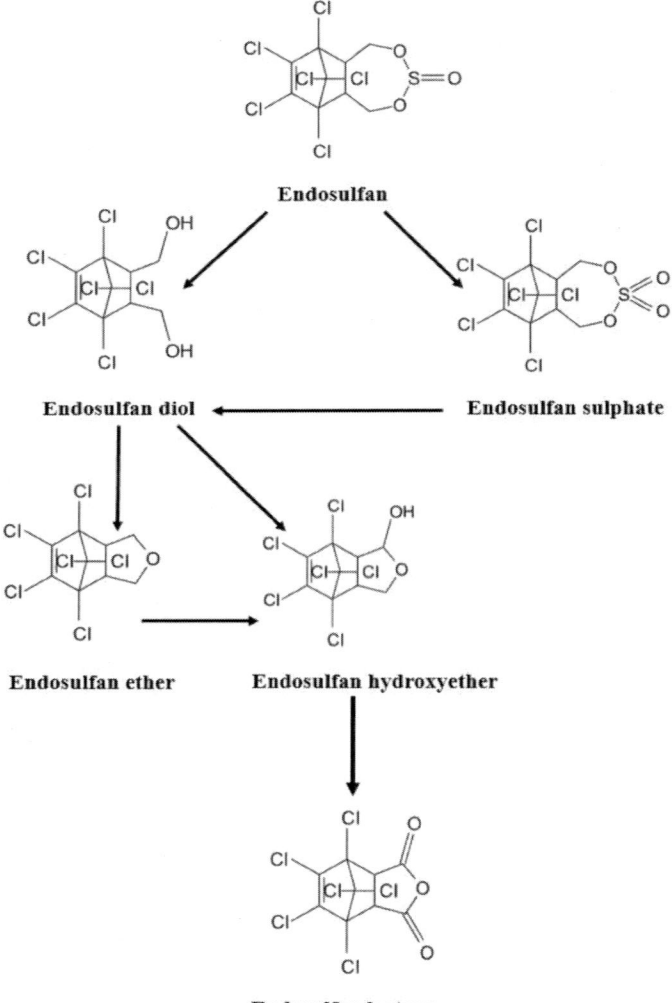

Figure 2. The schematic pathway of endosulfan degradation, as published by Shivamaria et al. [30].

2. RESIDUAL LEVELS OF ENDOSULFAN IN THE TROPICAL REGION

Residue concentrations of endosulfan are highest and most prevalent in or nearby regions with intense agricultural activity. Nevertheless, literature also revealed that the presence of endosulfan and their transformation products in other remote areas where agricultural activities cannot be found. The residues have been detected in a variety of media including air, soil, surface water, sediments, plants, aquatic vertebrates and invertebrates, terrestrial organisms, and in humans in the tropical region.

2.1. Air

Endosulfan contaminates air through direct application on crops and field dissipation and volatilization. A study carried out in high-altitude forests in the humid tropics, in Costa Rica revealed that annually averaged high air concentrations of endosulfan in the areas with intensive agricultural activities [32]. The parent compounds were much more abundant than the degradation product (endosulfan sulfate) in air samples and the most abundant pesticide was α-endosulfan, with concentrations ranging from 12 to 440 pg/m^3. This was observed in the densely populated central valley of Costa Rica, which is consistent with the usage of endosulfan in the coffee plantations that dominate the agricultural activity in that part of the country. Meire et al. [21] measured the pesticides residues in two mountain regions in tropical and temperate environments in Brazil. In this study, endosulfans were detected at all sampling sites exhibiting the highest air concentrations ranging from 43 to 5600(pg/m^3). In addition, the predominance of α-endosulfan over the total endosulfan profile (58 - 87%) was observed at all sites with extremely high air concentrations followed by β-endosulfan (8 - 27%) and endosulfan sulfate (2 - 16%). The seasonal variations further revealed that higher ratios of α-endosulfan/β-endosulfan were mainly observed during the summer (5-8) compared to winter period (2-5) indicating that more rapid or enhanced weathering of endosulfan during the summer period. Various studies [33, 34] indicate that β-endosulfan is degraded more quickly than α-endosulfan in the atmosphere. Therefore, α/β - endosulfan ratio in the atmosphere tends to increase with increasing distance from sources. The passive atmospheric sampling along the coastline of India also revealed high concentrations of α-endosulfan [33]. The sum total of endosulfan isomers and its metabolite (endosulfan sulfate) ranged from 0.45 - 1122 pg/m^3 along the coastline. The average isomeric ratio of endosulfan in the study was close to the technical grade and the concentration of the metabolite endosulfan sulfate was very low, indicating the ongoing application of technical endosulfan, especially in the cotton fields and tea plantations in India. Nevertheless, the National Institute for Occupational Safety and Health recommended a limit of 0.1 mg/m^3 for endosulfan in workplace air as an average over a period of 10 hours [35].

2.2. Soil and Sediments

Contamination of soil with endosulfan is primarily occurred by direct applications on crops to eliminate target pests. Dry or wet deposition on soil is also occurred by aerial spraying of endosulfan. The investigation of leaching of technical grade endosulfan in topsoil in a tropical environment of Brazil revealed the presence of two endosulfan isomers and endosulfan sulfate in percolated water indicating the degradation of the parent compound and leaching along vertical soil [36]. α-Endosulfan concentrations in water collected in the lysimeters varied from 0.01 to 0.20 μg/L in 17% of the samples, whereas β-endosulfan was detected in 33% in concentrations ranging from 0.01 to 0.38 μg/L. The metabolite endosulfan sulfate was also detected in 37% of the samples in higher concentrations, reaching 0.64 μg/L. Much higher adsorption coefficients (KD) of both endosulfan isomers in the deeper soil horizon were observed [36] and this indicates that molecules will probably be retained in the deeper layers (30-42 cm). The combined effect of clay and organic matter content may account for the higher retention of endosulfan isomers in the deep layers. A number of soil properties including organic matter content, soil texture, soil acidity, iron (Fe) and manganese (Mn) oxide and clay content can influence the retention and mobility of insecticide in a soil profile [37]. In addition, carbon sources present in natural soil and the concentration of endosulfan in the soil also influence the degradation of the parent compound. The results further indicate significantly higher field dissipation of endosulfan when compared to temperate regions. High air temperatures (monthly average temperatures from 22 to 27°C and the maximum of up to 38°C) could be attributed for the higher field dissipation. Another study verified that about 23.3% of the total applied amount of endosulfan on freshly tilled soils was lost by volatilization at temperatures between 20 to 30°C [38]. The characteristics of the pesticides themselves, the modes of application (especially whether they are applied subsurface or at the surface), and surface soil moisture status also affect the magnitude of these losses [38]. On the other hand, the addition of isolated bacterial cells belonging to *Bacillus* species into soil system contaminated with endosulfan causes an enhanced degradation of endosulfan isomers but the presence of sodium acetate or sodium succinate inhibited the degradation of endosulfan to different extents in soils inoculated with *Bacillus* species [39]. The rate of biodegradation progressed with the increase in endosulfan concentration up to 5.0 mg/g soil, followed by an inhibitory effect at higher concentrations, reaching a total loss of biodegradative activity at 10 mg/g soil. Ntow, [40] studied the dissipation of endosulfan in field grown tomato (*Lycopersicon esculentum*) and cropped soil at Akumadan in Ghana. A monophasic dissipation model in first order kinetics [41] was used in order to describe the dissipation of residues of endosulfan isomers in cropped soil. In tomato cropped soil, endosulfan followed essentially first order kinetics. Endosulfan concentration gradually decreased with time during the study period and the calculated DT50 and DT90 confirmed more or less similar persistence of α- and β- isomers of endosulfan in tomato cropped soil. Rosendhal et al., [14] studied the fate of pesticides in two representative horticultural soils (Acrisol and Arenosol) and plants (*Solanum macrocarpon* L.) after field application of endosulfan in Benin, West Africa. The results revealed that dissipation of endosulfan compounds faster for the Acrisol than for the Arenosol in this study. The

DT50 values indicated 4.8-13 days for α-endosulfan and 11-64 days for β-endosulfan. Endosulfan was, therefore, less strongly adsorbed and thus more readily available for dissipation in the Acrisol. A potential reason for the differences in dissipation rate between the two soils could be a poor microbial activity in the Arenosol, due to toxic levels of endosulfan residues in the soil.

The contamination of sediments occurs by atmospheric deposition and washouts of applied endosulfan on agro fields. Volta Lake, one of the largest man-made lakes in the world as well as the most important inland water resource in Ghana has also been contaminated with organochlorine pesticides. The residue levels of α, β isomers, and endosulfan sulfate were < LOQ - 0.89, < LOQ - 1.19 and 0.00 - 2.68 ng/g dw respectively [42]. Sediment contamination with endosulfan in two streams at Akumadan and Tono in Ghana was investigated [43]. In the study, endosulfan concentrations increased to 0.49 and 1.34 µg/kg in streambed sediment in the Akumadan stream at upstream and downstream sites, respectively in the dry season and in the wet season. The availability of endosulfan in sediments of Chantaburi estuary in Thailand was found after 3 years of its ban in the country [44]. The level of endosulfan (49.40 µg/kg) was more than two orders of magnitude higher [45] than that of Canadian [46] and USEPA marine sediment quality guidelines [47].

2.3. Water

Non-point source agricultural pollution is regarded as the greatest threat to the quality of surface waters. One of the most important routes leading to non-point source agricultural pollution of surface waters is a runoff. Endosulfan may enter the aquatic ecosystem through runoff, direct spray, leaching through the soil and volatilization into the atmosphere, and later as precipitation. In addition, the irrigation systems in modern agriculture have contributed to a more efficient spreading of pesticides like endosulfan [48] in surface water. In surface water of Akumadan and Tono streams in Ghana showed a relative downstream increase in the concentration of current-use pesticides including endosulfan after the runoff event. Vegetable fields dominated all of the downstream sub-catchments. On the contrary, runoff did not result in any significantly increased contamination in upstream sites where no vegetable fields were present. In general, endosulfan α-isomer is less water-soluble compared to the β-isomer (Table 1). Endosulfan sulfate is the most frequent form of the contaminant of endosulfan in river waters as faster biological degradation of the endosulfan isomers to endosulfan sulfate may occur in rivers in the presence of microbial activities [49]. A significant endosulfan sulfate production was observed from endosulfan formulation added to the artificial mesocosms. These mesocosms containing irrigation water obtained from the river were shown to be inhabited by a sizable population of micro- and meso-fauna (and, presumably, microflora) at the time of endosulfan treatments [49]. Nevertheless, endosulfan sulfate was not observed as a product of biological activities of algal blooms grown in a laboratory culture of incubated river water. A study carried out by Miles and Moy [50] stated that high populations of blue-green algae (cyanobacteria) failed to cause the significant formation of endosulfan sulfate. Nevertheless, conversions to endosulfan lactones are possible in river water [49]. In addition, the results showed that the endosulfan is stable at pH 5, with increasing rates of disappearance at alkaline pH values. The formation of endosulfan diol evidently occurs at alkaline pH values. In river water, blue-green algae and other organisms may raise pH and in turn, the chemical conversion yields the endosulfan diol in the medium. In surface water, endosulfan is usually attached to floating soil particles or can be found attached to sediments at the bottom. The small amounts of endosulfan that dissolve in the water break down over time. Spark and Swift, [51] stated that there is significant mobility of these chemicals through the soil for the presence of endosulfan in ground waters. In an agriculture field at Cote d'Ivoire in West Africa, endosulfan was detected in 85% of contaminated wells at rates exceeding the standards recommended for drinking water, of 0.1 μg/L. The average residue levels in all contaminated water in wells were 3.21 μg/L for α and 2.18 μg/L for β endosulfan while the maximum concentrations measured were 25.28 μg/L for α endosulfan and 13.74 μg/L for the β isomer [52]. The U.S. EPA recommends that the amount of endosulfan sulfate in lakes, rivers, and streams should not be more than 62 µg/L to prevent any harmful health effects from occurring in people who drink the water or eat fish that live in the water. In addition, the criterion established for the protection of aquatic life in freshwater includes total endosulfan of 60 ng/L for acute exposure and 3.0 ng/L for chronic exposure [53].

2.4. Lagoons and Oceans

The ultimate sink of most of the pollutants in the sediments of estuarine and oceans. The endosulfan is a high-risk pollutant in coastal aquatic environments due to its different transport routes including runoff and atmospheric deposition [54, 55]. Once they reach to coastal bodies, these pollutants including endosulfan deposit in sediments (via exchanging between the media of water and sediments) and organisms (via dietary, accidental or occupational routes). Although still, environment prevents the bringing back of endosulfan to the water column, dredging would potentially increase the dissolution of these pollutants in sediment to the water column. Ramirez-Elias et al., [56] studied the sediments of Sabancuy lagoon in tropical Mexico to determine the pesticide residues. The presence of endosulfan family showed significant differences ($P < 0.05$) among sampling sites and climatic periods. The levels were dominant in the dry season. Mean concentrations for α-endosulfan (0.642 ng/g), β-endosulfan (0.573 ng/g) and endosulfan sulfate (0.641 ng/g) were reported in sediments. On the contrary, studied from some tropical coastal areas recorded non-significant levels of endosulfan. The preliminary study of organochlorine pesticide residues on sediments of Bivalvia fishing ground at Eastern Part of Coastal Semarang, Indonesia showed that the average concentration of endosulfan was 7.1 µg/L in sediments [57]. In addition, Barasa, [58] investigated the levels of organochlorine pesticide residues on water, sediments, seaweeds, and fish from four sites (Sabki, Kilifi, Ramisi, Mombasa) along Kenyan coast between 1996 and 1997. Results revealed 0.15 - 0.19 µg/L endosulfan in all water samples, 9.22 - 26.9 µ g/L in all seaweed samples, 2.94 - 12.0 µg/L in all sediment samples and 10.4 - 40.2 µg/L in all fish samples during the wet season. The findings suggest that the coastal environment of Kenya is relatively unpolluted. Nevertheless, CCME [53] stated that the maximum levels of total endosulfan for the protection of aquatic life in seawater varies between 90 and 2.0 ng/L for acute and chronic exposure, respectively.

2.5. Biological Materials

2.5.1. Flora

The contamination of flora occurs either by direct application on foliage or plant uptake by roots. The uptake characteristics of endosulfan, including α-, β-isomers and sulfate-metabolites in carrot and potato plants that were grown in soils treated with endosulfan at concentrations of either 2 (low) or 10 (high) mg/kg soil was investigated [59]. Interestingly, the endosulfan uptake extent varied with the type of crop, endosulfan isomer, plant growth duration, and plant compartments. Although plant showed higher concentrations at the earlier period of the experiment, the concentrations decreased with time as observed from other similar studied carried out in tropics. Compared to the total endosulfan in carrot plants, potatoes showed higher concentrations. The uptake sequence of carrot plant was α-endosulfan > endosulfan sulfate > β endosulfan while the sequence in potato was endosulfan sulfate > α-endosulfan > β endosulfan. Endosulfan isomers absorbed via the roots of carrot were not transferred to shoots of carrots in low endosulfan soil. Chen et al., [60] stated that fine roots of plants are major pathways to return the absorbed hydrophobic organic chemicals to soils. In addition, Hwang et al., [59] stated that abundant fat-soluble vitamins contained in carrot roots could increase the retention capacity of endosulfan residues in root parts. On the contrary, transfer of root-absorbed endosulfan isomers transferred to shoot parts of carrots when excessive endosulfan residues were present. According to Anderson et al., [61], this transfer increases over time as the surface area of leaves growing bigger. In potatoes, the absorbed endosulfan isomers were more abundant in tuber parts while endosulfan sulfates were highest in roots, followed by tubers and shoots. The relatively higher mobility of endosulfan-sulfate from below ground to above ground parts than endosulfan isomers was confirmed by the presence of a large portion of the endosulfan-sulfate in shoots of carrots and potatoes. Most importantly, the edible parts of carrots and potatoes exceeded the MRL standards for endosulfan, 0.05 mg/kg plant [62]. On the other hand, α-endosulfan dissipated from plant surfaces, rapidly when compared to soil and almost completely (DT50: 1.6 h), whereas the initial fast decline of β-endosulfan was followed by a second phase of much slower dissipation, resulting in a half-life of 6.7 h. Rudel and Waymann [63] investigated the volatilization of endosulfan from plant surfaces in wind tunnel experiments at 20 to 25°C and reported that 60% of the applied amount of endosulfan volatilized from French beans within 24 hours. The distributions of endosulfan residues (α-, β-isomers, and sulfate-metabolite) in cucumbers grown in soils treated with endosulfan at concentrations of 20 and 40 mg/kg were assessed using indoor and outdoor experiments [64]. Overall, these results indicate that the most absorbed endosulfan residue in the cucumbers was endosulfan sulfate in the 20 mg/kg treated plots and the α-isomer in the 40 mg/kg treated plots while α-isomer was the most dominant isomer in the cucumber in the outdoor experiment. Singh and Singh [65] reported that the dominant endosulfan residue absorbed from soils by various plants was β-isomer, which was contrary to the findings of this study. In addition, the most endosulfan isomers in the outdoor test were found in the roots (> 80%), followed by the leaves (6.1 – 12.8%) and stems (2.2 – 8.0%); the uptake amount of endosulfan sulfate by the cucumber occurred in the following order: root (79.3 – 87.8%) > stems (11.2 – 12.2%) > leaves (0.0 – 9.5%).

2.5.2. Fauna

Contamination of endosulfan in fauna is primarily due to bioconcentration from surrounding media or biomagnification via food webs. A study carried out by OBEPAB in cotton producing areas of Central Bénin found α endosulfan residues in aquatic animal species in the rivers of Dridji, including *Clarias gariepinus* (fish), *Cardisoma armatum* (crab), *Bufo regularis* (toad) and *Xenopus muelleri* (frog), at levels as high as 75 ng/g [66]. The octanol-water partitioning coefficients (log Kow) are respectively 4.74 and 4.79 for α and β isomers (Table 1) indicating a high potential for bioaccumulation of these isomers in aquatic biota. The levels of organochlorine pesticides (OCPs) in water, sediment, Nile Tilapia and African Catfishes from the Volta Lake, Ghana were investigated in order to evaluate the suitability of the water and fish for continuous human consumption [42]. All isomers and metabolite of endosulfan were detected in all samples analyzed except α-endosulfan in water. Nevertheless, endosulfan concentration ranged from < LOQ to 1.43 ng/g for sediment and fish samples.

In addition, Venugopalan and Rajendran [67] studied the pesticides in three mollusks (the oyster *Crassostrea madrasensis* and the clams, *Meretrix casta* and *Katalysia opima*) collected from Vellar Estuary in the Southeast coast of India. All samples contaminated with mean endosulfan residues of 0.42 ng/g ww. The authors also studied the sensitivity of the bivalves and it was in the order of *C. madrasensis* > *K. opima* > *M. casta*. The author further detected the endosulfan in striped mullet *Mugil caphalus* L. (Perciformes: Mugilidae) and the cat fishes *Mystus gulio* Hamilton (Siluriformes: Bagrinae). Residue concentrations of endosulfan in these two species varied between 0.02 to 2.47ng/g w/w. Results indicated that *M.cephalus* was more sensitive than *M. gulio*. Rodriguez et al., [2] also recorded that aquatic organisms of commercial importance such as bivalves, crustaceans, and fish in the Gulf of Mexico and the Caribbean Sea had higher concentrations of β-endosulfan, followed by the α isomer and endosulfan sulfate indicating recent use of the pesticide. Rand et al., [68] conducted a comprehensive probabilistic aquatic ecological risk assessment to determine the potential risks of existing exposures to endosulfan and endosulfan sulfate in freshwaters of South Florida based on historical data (1992–2007). This included actual measured concentrations in surface freshwaters of 47 sites in South Florida from historical data to U.S.EPA numerical water quality criteria. Results showed several sites with endosulfan water quality violations including the southeastern boundary of Everglades National Park (ENP). Based on these results the second step of the study focused on the acute and chronic risks of endosulfan at nine sites by comparing distributions of surface water exposure concentrations of endosulfan [i.e., for total endosulfan (summation of concentrations of α- and β-isomers plus the sulfate), α- plus β-endosulfan, and endosulfan sulfate (alone)] with distributions of species effects from laboratory toxicity data. In the second step, the distribution of total endosulfan in fish tissue (whole body) from South Florida freshwaters was also used to determine the probability of exceeding a distribution of whole body residues of endosulfan producing mortality (critical lethal residues). The results indicated a potential chronic risk of 5% except for one particular site (9.2%) for total endosulfan among the selected 09 sites. The marsh killifish, flagfish, and mosquitofish showed the highest potential risk for lethal whole body tissue residues of endosulfan. The study thus reveals potential risks of total endosulfan to freshwater organisms in South Florida. The reference values for α and β endosulfan for the protection of freshwater aquatic life indicate that the acute and chronic effects are reported at concentrations between 220,000,000 and 56,000,000 ng/L, respectively. As for marine organisms, acute effects exist at 34,000,000 ng/L and chronic effects begin at concentrations of 8,700,000 ng/L [69].

3. HEALTH IMPACTS

The exposure to endosulfan on a human can be identified as dietary (via food or water), occupational (via skin) or accidental (via skin or inhalation). The warm tropical conditions coupled with socioeconomic and cultural aspects contribute to insufficient personal protective equipment being used in the application of pesticides, which increases risks to human health and exposure [70]. Human glial and neuronal cells are the most sensitive to endosulfan toxicity [71]. Endosulfan is a neurotoxicant since it binds to the gamma-aminobutyric acid (GABA)-gated chloride channel receptor inhibiting GABA-induced chloride flux across membranes in the central nervous system [2]. Humans may develop some neurological diseases induced by it, and nerve cells once damaged could not be regenerated [72].

In mammalian systems, endosulfan is metabolized to endosulfan sulfate, and also to endosulfan diol, which is further metabolized to endosulfan ether, hydroxy ether, and lactone. A survey carried out in 21 villages of the regions of Kita, Fana, and Koutiala, in Mali in 2001, found 73 pesticide poisoning cases and endosulfan was the main culprit [66]. In addition, another survey carried out during the period of 2003-2004 in Senegal, mainly in the region of Velingara (cotton growing area per excellence), identified endosulfan as the cause of 31.2 to 39.9% of the 162 poisoning cases, including 20 deaths. Most of the cases (73.2%) occurred during the endosulfan application [66]. The experimental results revealed that the exposure to sublethal doses of endosulfan, its isomers, and its metabolites induce DNA damage in Chinese hamster ovary (CHO) cells and human lymphocytes [73]. The study also revealed that, while the metabolites were more genotoxic than the parent compound in CHO cells, endosulfan displayed relatively strong DNA-damaging activity in human lymphocytes. Moreover, the study demonstrated that α-endosulfan produces slightly more DNA strand breaks than the β-isomer, both in CHO cells and in human lymphocytes. The effect of endosulfan on the male reproductive system was studied by selecting children in the village of Kasargod in Kerala, India who had been exposed to endosulfan during the 20 years of aerial spraying of cashew nut plantation [74]. Although Environmental exposure to a single chemical over a long period of time is very rare, the researchers came across a situation where endosulfan was the only pesticide that had been aerially sprayed two to three times a year for more than 20 years. The study parameters included recording of clinical history, physical examination, sexual maturity rating (SMR) according to Tanner stages, and estimation of serum levels of testosterone, luteinizing hormone (LH), follicle-stimulating hormone, and endosulfan residues. The results of the study suggested that endosulfan exposure in male children may delay sexual maturity and interfere with sex hormone synthesis. In addition, Rodríguez et al., [2] stated that in the Kerala state, adverse effects on human health and calves with congenital deformities are being detected since 1979. During the 1990s, an unusual increase in health problems was reported in Kasaragod, one district of Kerala. In 2001, more cases of children with hydrocephalus, cerebral palsy, blindness, deformed hands, and chronic skin problems were reported [75]. However, the Agency for Toxic Substances and Disease Registry (ATSDR) identifies people with liver or kidney disease; pre-existing anemia or hematological disorders; neurological problems especially seizure disorders; people with HIV/AIDs and people with protein-deficient diets such as the malnourished poor, chronic alcoholics and dieters as vulnerable groups [76]. In addition, unborn children, infants and the elderly are some populations particularly sensitive to endosulfan's neurotoxic effects.

4. TOXICITY ON LABORATORY ANIMALS

Data from laboratory animal studies are often used to draw inferences about the potential hazards of pesticides when no adequate data are available. Investigations of the toxic effects of prolonged exposure to endosulfan on some blood parameters in sexually matured male albino rats revealed that the total counts of erythrocytes and hemoglobin were decreased and leucocytes were increased. A significant increase in basophils and monocytes was observed through differential counts. The levels of leucocytes serum glucose, urea, creatinine, and bilirubin increased significantly, suggesting that the synthetic endosulfan had remarkable toxic effects on the hematological and biochemical parameters in these experimental rats [77]. Endosulfan has a relatively high potential to bioaccumulate in fish. The continuous exposure of three freshwater catfishes; *Heteropneustes fossilis*, *Mystus cavasius* and *M. vittatus* at 28 ± 2°C to 0.0005 mg/L of endosulfan for 96 h were done to investigate the deposition and metabolism of endosulfan [78]. After 96 h the fish were killed by decapitation, the tissues, brain, gill filaments, gut, gall bladder, kidney, liver, and muscle were immediately dissected and weighed. Endosulfan sulfate was the principal metabolite in different tissues of the three catfishes studied as observed in mammalian systems. Nevertheless, endosulfan sulfate formation is not a detoxification product as it is as toxic as the technical material [79]. The principal detoxification products were endosulfan alcohol in *H. jossilis*, endosulfan lactone, endosulfan hydroxy ether and endosulfan ether in *M. cavasius* and endosulfan alcohol and endosulfan ether in *M. vittatu*. In all the three catfish's liver and kidney were the principal sites of storage. Toxicity tests of the mangrove guppy *Gambusia puncticulata* Poey, a common fish in Jamaican mangrove environments revealed that the 24hr LC50 for endosulfan was 0.050 µg/L, concentrations of 0.1 µg/L resulted in an instant reaction in the fish and 100% mortality within 1 hour [80]. The ecotoxicity of sub-lethal concentrations of endosulfan on Tilapia and its impact on the reproductive physiology of the fish was investigated [81]. The 28-day sub-lethal toxicity of endosulfan to Tilapia (9 - 14 cm long: 20 - 40g) was assayed under static conditions in continually aerated glass aquaria containing 20 L of dechlorinated water and 15 fish at 27 - 28°C. The fish were assayed in triplicates with the concentrations of 0.0005, 0.0075, 0.008, 0.009 and 0.01 endosulfan. Symptoms such as darting, shuddering, side-swimming, and death were observed over time. These observed results were categorized into different toxicity concentrations: the no-observable-toxic-effect-concentration (NOTEC), least-observable toxic-effect-concentration (LOTEC), pronounced-observable-toxic-effect-concentration (PROTEC) and least-acute-toxic-effect-concentration (LATEC). At the end of the experimental period of 28 days, three fish from each replicate were removed and assayed. Whole body parts or gonads were cut into small parts and extracted with a chloroform-methanol mixture. The toxicity data suggest that 0.0025 - 0.05 mg/L may be defined as NOTEC, the 0.008 mg/L dose LOTEC, 0.009 mg/L PROTEC. The freshwater prawn *Macrobrachium dayanum* were subjected to acute (96 hr) static bioassay toxicity [82]. The LC50 values obtained were 0.006 mg/L and the prawns were then exposed for 30 days under the static condition to their sub-lethal concentrations viz. 0.003 mg/L, 0.002 mg/L and 0.001 mg/L of endosulfan. The results revealed decreased growth in direct relation to the increase in the concentrations of the toxicants. The impact of the insecticide endosulfan was also assessed on the growth of the African giant snails, litter living animals found in cocoa fields throughout tropical Africa [83]. Two doses of endosulfan, C1, 6.25 g/L, and C2, 12.50 g/L were applied twice with one month of the interval to the litter of the snails. The daily weight gain (g/j) of - 0.028 ± 0.004 for C1 and - 0.033 ± 0.007 for C2 showed that snails which received endosulfan in their feed had a very weak growth compared to the control one, which has a daily weight gain of 0.032 ± 0.006.

5. Impacts on the Tropical Environment

The endosulfan courses great threat to the balance of the environment. In agroecosystems essentially, earthworms are considered most beneficial because of their contribution to complex processes such as litter decomposition, nutrient cycling and soil formation [84]. Dasgupta et al., [85] reported that growth, reproduction, and respiration of the tropical epigeic earthworm Perionyx excavatus were affected by sub-lethal doses of endosulfan, which were much lower than the recommended agricultural dose (RAD). The cocoon and juvenile production were significantly reduced by the insecticide even at the lowest dose considered in the study. Similarly, soil microorganisms form a vital part of the soil food web [86] and microbial biomass serves as a potential measure of ecosystem functioning [87]. These organisms participate in oxidation, nitrification, ammonification, nitrogen fixation, and other processes which lead to decomposition of soil organic matter and transformation of nutrients [88], they can also store C and nutrients in their biomass which are mineralized after cell death by surviving microbes [89] Soil Dehydrogenase is an indicator of overall microbial activity, as it occurs intracellularly in all living microbial cells and is linked with microbial oxydo-reduction processes [90]. The increased dehydrogenase and other parameters including fluorescein diacetate hydrolase, acid phosphatase, aryl sulphatase activities and microbial biomass of the soil with 1 mg/L endosulfan indicated the possible involvement of soil microorganisms and their enzymes in degradation of endosulfan [91]. Nevertheless, significantly decreased soil nitrogenase activity suggested that endosulfan or its metabolites may pose a toxicological threat to nitrogen fixers in the soil. The effect of endosulfan and some other pesticides on microbial respiration parameters in soil from cultivated and fallow plots from Burkina Faso, Africa revealed that repeated application of endosulfan in the field affected soil microorganisms' respiration parameters [92]. Basal respiration which is believed to reflect the potential of the microbial activity showed a tendency to be higher in the cultivated soil as compared to fallow. Basal respiration and lag time could be useful tools to discriminate soil quality in semi-arid condition and to assess the impact of pesticides on soil biology. The application of endosulfan caused to decrease basal respiration in this study.

6. SUMMARY AND CONCLUSION

The literature reveals that application of endosulfan on target crops contaminates different compartments in the environment. Considering the residue levels of endosulfan in the atmosphere, tropical regions considered in the present chapter showed higher levels when compared to alpines, temperate and sub-tropical regions. High alpine stations in Switzerland, Austria, and Germany showed 10.5 - 13.7 pg/m^3 of α- isomers and 0.43 - 0.62 pg/m^3 of β- isomers [93] while temperate air in Italy indicated the range of 30 - 70 pg/m^3 in rural and agricultural regions [94]. In Dalian in Northeast China recorded 0.1 - 52.6 pg/m^3 of α-endosulfan [95]. The higher residue levels in tropical air may be due to fresh applications. Once the endosulfan released, the degradation proceeds via abiotically and biotically in the soil. α and β isomers and endosulfan sulfate are the major products during degradation. Nevertheless, the most toxic endosulfan sulfate persists in soil owing to its higher Kd. On the other hand, higher dissipation of endosulfan, its isomers, and metabolites from tropical topsoil is characterized by the higher temperatures compared to temperate regions. The characteristics of heavy precipitations as rain influence the offsite movement of these isomers and metabolites in addition to vertical movements in the soil. Moreover, being the final destination of most of the pollutants, marine environment possesses higher levels of endosulfan residues. The recent studies from tropical marine environment greatly witness for the presence of higher endosulfan residues in different biotic samples. Recent investigations also reveal that the levels of endosulfans are still higher in certain compartments of the temperate regions although many countries in the region have banned the use of endosulfan in the 1990s. The-α, β- endosulfan, endosulfan sulfate in water, sediment and soil samples in agricultural farms in the Okanagan Valley in British Columbia in Canada approached the guidelines for the protection of aquatic life [96] confirming that the significant potential of cold trapping of endosulfan in such environments. Similarly, the occurrence and distribution of endosulfan in water, sediments and fish tissues in protected lands in the tropical environment of South Florida recorded significantly elevated endosulfan concentrations Quinete et al. [97]. The endosulfan concentrations observed in surface water and sediments were 158 ng/L and 57 ng/L respectively. In addition, the presence of elevated levels of endosulfan sulfate (up to 371 ng/g) in whole fish tissue was also observed from tropical South Florida [97]. Laboratory ecotoxicity studies using animals at tropical temperatures depending on the administrative doses prove that endosulfan has the capability to concentrate its sulfate metabolite and isomers in some organs with high concentrations. In fish, liver and kidney are the target organs. The parent compound and its transformation products affect the reproductive physiology of many kinds of organisms including fish, mammals, and even humans.

The findings of this scientific literature survey conclude that although the levels of endosulfan and its transformation products in tropical environment thought to be lower than the temperate, sub-tropical or alpine regions, the illegal or continuous use of endosulfan in tropical regions silently bring significant levels to the environment which can ultimately pose threats to both environment and human health. The study thus proposes the urgent need for environmentally friendly alternatives to eliminate the negative impacts of this pesticide or stop its future use completely.

REFERENCES

[1] Maier-Bode, H. 1968. "Properties, effect, residues and analytics of the insecticide endosulfan". *Residue Reviews*, 22, 1 - 44.

[2] Kumar, M. and Philip, L. (2006). "Enrichment and isolation of a mixed bacterial culture for complete mineralization of endosulfan". *Journal of Environmental Science and Health Part B*, *41*, 81–96.

[3] Rodríguez, G. N., Sánchez, C. L., Estrada, A. S., Chavez, M. D. R. C., Reynoso, F. L., Vazquez, A. P. and Iourii Nikolskii, G. (2016). "Endosulfan: Its Isomers and Metabolites in Commercially Aquatic Organisms from the Gulf of Mexico and the Caribbean". *Journal of Agricultural Science*, *8*(1) Accessed November 29, 2018. doi:10.5539/jas.v8n1p8.

[4] Extoxnet. (1996). EXTOXNET (Extension Toxicology Network), "Pesticide Information Profiles: Endosulfan". http://extoxnet.orst.edu/faqs/, Accessed December 12, 2018.

[5] Weber, J., Halsall, C. J., Muir, D., Teixeira, C., Small, J., Solomon, K., Hermanson, M., Hung, H. and Bidleman, T. (2010). "Endosulfan, a global pesticide: A review of its fate in the environment and occurrence in the Arctic". *Science of the Total Environment*, *408*, 2966–84.

[6] Shen, L., Wania, F., Ying, D. L., Teixeira, C., Muir, D. C. G. and Bidleman, T. F. (2005. "Atmospheric distribution and long-range transport behaviour of organochlorine pesticides in North America". *Environmental Science and Technology*, *39*, 409–20.

[7] US EPA. (2002). *Reregistration Eligibility Decision for Endosulfan.* EPA 738-R-02-013. Pollution, Pesticides and Toxic Substances (7508C), United States Environmental Protection Agency. https://nepis.epa.gov/Exe/ZyPURL.cgi?Donkey=P1007KRN.TXT Accessed December 12, 2018.

[8] Soto, A. M., Chung, K. L. and Sonnenschein, C. (1994). "The pesticides endosulfan, toxaphene, and dieldrin have estrogenic effects on human estrogen-sensitive cells". *Environmental Health Perspectives*, *102*(4), 380-383.

[9] Daniel, C. S., Agarwal, S. and Agarwal, S. S. (1986). "Human red blood cell membrane damage by endosulfan". *Toxicology Letters*, *32*(1-2), 113-118.

[10] PANAP. (2009). "Endosulfan. Pesticide Action Network Asia & Pacific (2nd ed.)". http://www.pan-germany.org/download/ Endo_09_ PANAP_monograph_2nd%20Edition. Accessed December 12, 2018.

[11] Kannan, K., Holcombe, R. F., Jain, S. K., Alvarez-Hernandez, X., Chervenak, R., Wolf, R. E. and Glass, J. (2000). "Evidence for the induction of apoptosis by endosulfan in a human T-cell leukemic line". *Molecular and Cellular Biochemistry*, *205*(1- 2), 53-66. Doi: 10.1023/A:1007080910396.

[12] Lu, Y., Morimoto, K., Takeshita, T., Takeuchi, T. and Saito, T. (2000). "Genotoxic effects of α-Endosulfan and β-endosulfan on human HepG2 cells". *Environmental Health Perspectives.*, *108*(6), 559-61. Accessed December 12, 2018. doi: 10.2307/3454619.

[13] Kelly, B. C. and Gobas, F. A. P. C. (2003). "An Arctic terrestrial food-chain bioaccumulation model forpersistent organic pollutants". *Environmental Science and Technology*, *37*(13), 2966-74.

[14] Rosendahl, I., Laabs, V., Atcha-Ahowe, C., James, B. and Amelung, W. (2009). "Insecticide dissipation from soil and plant surfaces in tropical horticulture of southern Benin, West Africa". *Journal of Environmental Monitoring*, *11*(6), 1157-1164. Accessed November 28, 2018. doi: 10.1039/b903470f.

[15] Ozer, S. (2005). "Measurement of Henry's law constant of organochlorine pesticides". MSc Diss., Izmir Institute of Technology, Turkey.

[16] Shen, L. and Wania, F. (2005). "Compilation, evaluation, and selection of physical–chemical property data for organochlorine pesticides". *Journal of Chemical and Engineering Data*, *50*, 742–768.

[17] IPEN. (2009). "Endosulfan in West Africa: Adverse Effects, its Banning, and Alternatives"1-31. https://ipen.org/documents/ endosulfan-west-africa, Accessed November 17, 2018.

[18] Peterson, S. M. and Batley, G. E. (1993). "The fate of endosulfan in aquatic ecosystems". *Environmental Pollution*, *82*, 143–52.

[19] Stockholm Convention. (2009). "Endosulfan: Canada's Submission of Information Specified in Annex E of the Stockholm Convention Pursuant to Article 8 of the Convention". http://chm.pops.int/

Portals/0/docs/Responses_on_Annex_E_information_for_endosulfan/Canada_090110_SubmissionEndosulfanInformation.doc, Accessed November 16, 2018.

[20] Stockholm Convention. (2017). Register of Specific Exemptions: Technical Endosulfan and Its Related Isomers. http://chm.pops.int/Implementation/Exemptions/SpecificExemptions/ Technical endosulfanRoSE/tabid/5037/Default.aspx, Accessed November 16, 2018.

[21] Meire, R. O., Lee, S. C., Yao, Y., Targino, A. C., Torres, J. P. M. and Harner, T. (2018). "Seasonal and altitudinal variations of legacy and current-use pesticides in the Brazilian tropical and subtropical mountains". *Atmospheric Environment*, 59, 108-116. Accessed December 13, 2018. https:// www.researchgate.net/ publication/ 235221197.

[22] Lee, J. B., Sohn, H. Y., Shin, K. S., Jo, M. S., Kim, J. E., Lee, S. W., Shin, J. W., Kum, E. J. and Kwon, G. S. (2006). 'Isolation of a soil bacterium capable of biodegradation and detoxification of endosulfan and endosulfan sulfate". *Journal of Agricultural and Food Chemistry*, 15, 54(23), 8824-8828.

[23] Schmidt, W. F., Bilboulian, S., Rice, C. P., Fettinger, J. C., McConnell, L. L. and Hapeman, C. J. (2001). "Thermodynamic, Spectroscopic, and Computational Evidence for the Irreversible Conversion of β- to α-Endosulfan". *Journal of Agricultural and Food Chemistry*, 49(11), 5372-5376.

[24] Sutherland, T. D., Home, I., Russel, R. J. and Oakeshott, J. G. (2002). "Gene cloning and molecular characterization of a two-enzyme system catalyzing the oxidative detoxification of beta-endosulfan". *Applied and Environmental Microbiology*, 68(12), 6237-6245.

[25] Kumar, K., Devi, S. S., Krishnamurthi, K., Kanade, G. S. and Chakrabarti, T. (2007). "Enrichment and isolation of endosulfan degrading and detoxifying bacteria". *Chemosphere*, 68(2), 317-322.

[26] Hussain, S., Arshad, M., Saleem, M. and Khalid, A. (2007). "Biodegradation of alpha- and beta-endosulfan by soil bacteria". *Biodegradation*, 18(6), 731-740.

[27] Lee, S. E., Kim, J. S., Kennedy, I. R., Park, J. W., Know, G. S., Koh, S. C. and Kim, J. E. (2003). "Biotranformation of an organochlorine insecticide, endosulfan by *Anabaena* species" *Journal of Agricultural and Food Chemistry*, 26, 51(5), 1336-13440.

[28] Awasthi, N., Singh, A. K., Jain, R. K., Khangarot, B. S. and Kumar, A. (2003). "Degradation and detoxification of endosulfan isomers by a defined co-culture of two Bacillus strains". *Applied Microbiology and Biotechnology*, 62(2-3), 279-283.

[29] Walse, S. S., Scott, G. I. and Ferry, J. L. (2003). "Sterioselective degradation of aqueous endosulfan in modular estuarine mesocosms: formation of endosulfan gamma-hydroxycarboxylate", *Journal of Environmental Monitoring*, 5(3), 373-379.

[30] Shivaramaiah, H. M., Sanchez-Bayo, F., Al-Rifal, J. and Kenndey, I. R. (2005). "The Fate of Endosulfan in Water". *Journal of Environmental Science and Health, Part B: Pesticides, Food Contaminants, and Agricultural Wastes*, 40, 5, 711-720.

[31] *Damalas, C. A. and Eleftherohorinos, I. G. (2011).* "Pesticide Exposure, Safety Issues, and Risk Assessment Indicators". *International Journal of Environmental Research and Public Health., 8 (12), 1402–1419.*

[32] Shunthirasingham, C., Gouin, T. D., Lei, Y. D., Ruepert, C., Castillo, L. E. and Wania, F. (2011). "Current use pesticide transport to Costa-Rica's high altitude tropical cloud forest". *Environmental Toxicology and Chemistry*, 30(12), 2709–2717.

[33] Zhang, G., Chakraborty, P., Li, J., Sampathkumar, P., Balasubramanian, K., Kathiresan, K., Takahashi, S., Subramanian, A., Tanabe, S. and Jones, K. C. (2008). "Passive atmospheric sampling of organochlorine pesticides, polychlorinated biphenyls, and polybrominated diphenyls ethers in urban, rural, and wetland sites along the coastal length of India". *Environmental Science and Technology*, 42, 8218–8223.

[34] Shunthirasingham, C., Oyiliagu, C. E., Cao, X., Gouin, T., Wania, F., Lee, S. C., Pozo, K., Harner, T. and Muir, D. C. G. (2010). "Spatial

and temporal pattern of pesticides in the global atmosphere". *Journal of mental Monitoring*, *12*, 1650–1657.

[35] ATSDR. (2015). Toxicological profile for endosulfan. Agency for Toxic Substances and Disease Registry. Accessed December 14, 2108. https://www.atsdr.cdc.gov/toxprofiles/tp41.pdf.

[36] Dores, E. F., Spadotto, C. A., Weber, O. L., Dalla, Villa, R., Vecchiato, A. B. and Pinto, A. A. (2016). "Environmental Behavior of Chlorpyrifos and Endosulfan in a Tropical Soil in Central Brazil". *Journal of Agricultural and Food Chemistry*, *25*, 64(20), 3942-3948.

[37] Mooncake, D. M. and Johnson, H. W. (1982). Tal bat and Bank. "Microbial metabolism and enzymology of selected pesticides". In: *Biodegradation and Detoxification of Environmental Pollutants*, edited by Chakrabarty, A.M., 1-32. CRC, Press Boca Raton, FL.

[38] Rice, C. P., Nochetto, C. B. and Zara, P. (2002). "Volatilization of trifluralin, atrazine, metolachlor, chlorpyrifos, α-endosulfan, and β-endosulfan from freshly tilled soil". *Journal of Agricultural and Food Chemistry.*, *50*, 4009−4017.

[39] Awasthi, N., Ahuja, R. and Kumar, A. (2000). "Factors influencing the degradation of soil-applied endosulfan isomers". *Soil Biology and Biochemistry*, *32*, 1697-1705.

[40] Ntow, W. J., Ameyibor, J., Kelderman, P., Drechsel, P. and Gijzen, H. J. (2007). "Dissipation of Endosulfan in Field-Grown Tomato (*Lycopersicon esculentum*) and Cropped Soil at Akumadan, Ghana". *Journal of Agricultural and Food Chemistry*, *55*, 10864–10871

[41] Morrica, P., Fidente, P., Seccia, S. and Ventriglia, M. (2002). "Degradation of imazosulfuron in different Soils-HPLC determination". *Biomedical Chromatography*, *16*, 489-494.

[42] Gbeddy, G., Glover, E., Doyi, I., Frimpong, S. and Doamekpor, L. (2015). "Assessment of Organochlorine Pesticides in Water, Sediment, African Cat fish and Nile tilapia, Consumer Exposure and Human Health Implications, Volta Lake, Ghana". *Journal of Environmental and Analytical Toxicology*, *5*, 297. Accessed November 23, 2018. doi:10.4172/2161- 0525.1000297.

[43] Ntow, W. J., Drechsel, P., Botwe, B. O., Kelderman, P. and Gijzen, H. J. (2008). "The impact of agricultural runoff on the quality of two streams in vegetable farm areas in Ghana". *Journal of Environmental Quality*, 37, 696-703.

[44] Sumith, J. A., Parkpian, P. and Leadprathom, N. (2009). "Dredging influenced sediment toxicity of endosulfan and lindane on black tiger shrimp (*Penaeus monodon* Fabricius) in Chantaburi River estuary in Thailand". *International Journal of Sediment Research*, 24(24), 455-464. doi: 10.1016/S1001-6279(10)60017-0.

[45] Fucungkoon, N. (1991). "Pesticide residue level of organochlorine in soil, water and sediment from various type of land use of Chantaburi River Basin". MSc Diss., Kasetsart University, Bangkok, Thailand.

[46] Environment Canada. (2002). "Canadian Environmental Quality Guidelines", Update 2002. http://ceqg-rcqe.ccme.ca/en/index.html.

[47] USEPA. (1996). "Water Quality Regulations", Delaware River Basin Commission, West Trenton, New Jersey. Administration Manual-Part III, 104.

[48] Galaviz-Villa, I., Landeros-Sánchez., C., Castañeda-Chávez, M. R., Martínez-Dávila, J. P., Pérez-Vázquez, A., Nikolskii-Gavrilov, I. and Lango-Reynoso, F. (2010). "Agricultural contamination of subterranean water with nitrates and nitrites: an environmental and public health problem". *Journal of Agricultural Science*, 2(2), 17-30. Accessed November 20, 2018.http://dx.doi.org/10.5539/jas.v2n2p17.

[49] Shivamaiah, H. M., Sanchez-Bayo, F, Al-Rifai, J. and Kennedy, I. R. (2005). "The fate of endosulfan in Water". *Journal of Environmental Science and Health, Part B: Pesticides, Food Contaminants, and Agricultural Wastes*, 40, 5, 711-720, doi: 10.1080/03601230500189311.

[50] Miles, J. R. W. and Moy, P. (1979). "Degradation of endosulfan and its metabolites by a mixed culture of soil microorganisms". *Bulletin of Environmental Contamination and Toxicology*, 23, 13–19.

[51] Spark, K. M. and Swift, R. S. (2002). "Effect of soil composition and dissolved organic matter on pesticide sorption". *Science of the Total Environment.*, 298, 147–161.

[52] Traore, S. K., Mamadou, K., Debmble, A., Lafrance, P., Mazellier, P. and Houenou, P. (2006). "Contamination of groundwater by pesticides in agricultural areas in *Côte d'Ivoire* (central, *south* and *southwest*)". *African Journal of Environmental Science and Technology*, *1*, 1-9.

[53] CCME (Canadian Council of Ministers of the Environment). (2014). "Canadian Environmental Quality Guidelines Summary Table". Water Quality Guidelines for the Protection of Aquatic Life. Accessed November 20, 2018. http://ceqg-rcqe.ccme.ca.

[54] Albert, L. A. and Benítez, J. A. (1996). "environmental impact of pesticides on coastal ecosystems", In A. V. Botello, J. L. Rojas-Galaviz, J. A. Benítez, and D. Zárate-Lomelí (Eds.), Gulf of Maxico. *Pollution and Environmental Impact: Diagnosis and Trends*. Mexico D. F.: EPOMEX, University of Campeche.107-123.

[55] Albert, L. A. (2014). "Pesticides and their risks to the environment". In A. V. Botello, J. Rendón von Osten, J. Benítez, & G. Gold-Bouchot (Eds.), Golfo de México. *Pollution and Environmental Impact: Diagnosis and Trends* (3rd ed., pp. 183-212). Mexico D. F.: UAC, UNAM-ICMyL, Merida Unit CINVESTAV.

[56] Ramirez -Elias, M. A., Cordova-Quiroz, A. V., Ceron-Breton, J. G., Ceron-Breton, R. M., Rendon-von O s-ten, J. and Cortes -Simon, J. H. (2016). "Dichloro -Diphenyl -Trichloroethane (DDT) and Endosulfan in Sediments of Sabancuy Lagoon, Campeche, Mexico". *Open Journal of Ecology*, *6*, 22-31. Accessed November 05, 2018. http://dx.doi.org/10.4236/oje.2016.61003.

[57] Suryono, C. A., Subagyo, Setyati, W. A., Susilo, E. S., Rochaddi, B. and Mahendrajaya, R. T. (2018). "The Preliminary Study of Organochlorine Pesticide Residues on Sediments of Bivalvia Fishing Ground at Eastern Part of Coastal Semarang". *IOP Conf. Series: Earth and Environmental Science*, *116*, 012093 Accessed November 29, 2018. doi :10.1088/1755-1315/116/1/012093.

[58] Barasa, M. W. (1998). "Studies on distribution of organochlorine residues in a tropical marine environment along the Kenyan coast". MSc Diss., University of Nirobi.

[59] Hwang, J. I., Zimmerman, A. R. and Kim, J. E. (2018). "Bioconcentration factor-based management of soil pesticide residues: Endosulfan uptake by carrot and potato plants". *Science of the Total Environment*, *627*, 514–522.

[60] Chen, Z. X., Ni, H. G., Jing, X., Chang, W. J., Sun, J. L. and Zeng, H. (2015). "Plant uptake, translocation, and return of polycyclic aromatic hydrocarbons via fine root branch orders in a subtropical forest ecosystem". *Chemosphere*, *131*, 192–200.

[61] Anderson, J. J., Bookhart, S. W., Clark, J. M., Jernberg, K. M., Kingston, C. K., Snyder, N., Wallick, K. and Watson, L. J. (2013). "Uptake of cyantraniliprole into tomato fruit and foliage under hydroponic condition: application to calibration of a plant/soil uptake model". *Journal of Agricultural and Food Chemistry.*, *61*, 9027–9035.

[62] CODEX Alimentarius. (2017). *"International Food Standards"*. http://www.fao.org/fao-whocodexalimentarius/standards/ pestres/ pesticides/en/, Accessed November 16, 2018.

[63] Rudel, H. and Waymann, B. (1992). "Volatility Testing of Pesticides in a Wind Tunnel", Proceedings of the Brighton Crop Protection Conference 1992, *Pests and Diseases*, *2*, 841-846.

[64] Hwang, J. I., Lee, S. E. and Kim, J. E. (2015). "Plant Uptake and Distribution of Endosulfan and Its Sulfate Metabolite Persisted in Soil". *PLOS ONE*, *10*(11), e0141728.Accessed December 15, 2018. https://doi.org/10.1371/journal.pone.0141728.

[65] Singh, V. and Singh, N. (2014). "Uptake and accumulation of endosulfan isomers and its metabolite endosulfan sulfate in naturally growing plants of contaminated area". *Ecotoxicology and Environmental Safety*, 2014, *104*, 189–193.

[66] Glin, L. J., Kuiseau, J., Thiam, A., Vodouhe, D. S., Dinham, B. and Ferrigno, S. (2006). *"Living with Poison: Problems of Endosulfan in West Africa Cotton Growing Systems"*. Pesticide Action Network UK, London.

[67] Venugopalan, V. K. and Rajendran, N. (1984). *"Pesticide pollution effects on marine and estuarine resources"*. DAE Research Project Report, Parangippettai. India: Centre for Advanced Study in Marine Biology, Annamalai University, 1- 316.

[68] Rand, G. M., Carriger, J. F. and Gardinali, P. R. (2010). "Endosulfan and its metabolite, endosulfan sulfate, in freshwater ecosystems of South Florida: a probabilistic aquatic ecological risk assessment". *Ecotoxicology, 19*, 879–900.

[69] EPA. (2014). *"National Recommended Water Quality Criteria, Aquatic Life Criteria Table: Alpha + beta Endosulfan"*. Accessed December 15m 2018. http://water.epa.gov/scitech/ swguidance/ standards/ criteria/current/index.cfm#organoleptic.

[70] Feola, G. and Binder, C. R. (2010). "Why don't pesticide applicators protect themselves? Exploring the use of personal protective equipment among Colombian smallholders". *International Journal of Occupational and Environmental Health, 16*, 11−23.

[71] Khan, K. H. (2012). "Impact of endosulfan on living beings". *International Journal of Biosciences (IJB), 2* (1), 9-17.

[72] Chan, M. P. L., Morisawa, S., Nakayama, A. and Yoneda, M. (2007). "Evaluation of health risk due to the exposure to endosulfan in the environment". AATEX 14, Special Issue, *Proc. 6th World Congress on Alternatives and Animal Use in the Life Sciences*. Tokyo, Japan, August. 543-548.

[73] Bajpayee M., Pandey, A. K., Zaidi, S., Musarrat, J., Parmar, D., Mathur, N. and Dhawan, A. (2006). "DNA damage and mutagenicity induced by endosulfan and its metabolites". *Environmental and Molecular Mutagenesis, 47*(9), 682-692. Accessed November 14, 2018. http://dx.doi.org/10.1002/em.20255.

[74] Saiyed, H., Dewan, A., Bhatnagar, V., Shenoy, U., She noy, R., Rajmohan, H. and Lakkad, B. (2003). "Effect of endosulfan on male reproductive development". *Environmental Health Perspectives, 111*(16), 1958-1962.

[75] Bejarano, G. F. (2008). "Endosulfan a threat to health and the environment". In F. Bejarano-González, J.Souza-Casadinho, J. M.

Weber, C. Guadarrama-Zugastu, E. Escamilla-Prado, B. Beristáin-Ruiz, F.Ramírez-Muñoz (Eds.), Endosulfan and its Alternatives in Latin America Mexico D. F.: *IPEN, RAP-ALRAPAM, UACH.*, 9-34.

[76] ATSDR. (2000). *"Toxicological Profile for Endosulfan"*. Agency of Toxic Substances and Disease Registry, Atlanta, USA. http://www.atsdr.cdc.gov/toxprofiles/tp41.html Accessed November 23, 2018.

[77] Das, B., Pervin, K., Roy, A.K., Ferdousi, Z. and Saha, A. K. (2010). "Toxic effects of prolonged endosulfan exposure on some blood parameters in albino rat". *Journal of Life and Earth Science*, *5*, 29-32.

[78] Rao, D. M. R. and Murty, A. S. (1982). "Toxicity and metabolism of endosulfan in three fresh water cat fishes". *Environmental Pollution (Series A)*, *27*, 223-231.

[79] Barnes, W. W. and Ware, G. W. (1965). "The absorption and metabolism of 14C-labelled endosulfan in the house-fly". *Journal of Economic Entomology*, *58*, 286-91.

[80] Williams, L. A. D. and Chow, B. A. (1993). "Toxicity of endosulfan to the mangrove swamp guppy *Gambosia Puncticulata* Poey and its ecological implication in Jamaica". *Philippine Journal of Science*, *122*, 323-328.

[81] Williams, W. O., Golden, K. and Mansingh, A. (1999). "Ecotoxicity of insecticide residues in Jamaica: Sub lethal toxicity levels of endosulfan on the fish Tilapia and its effects on lipid content of gonads". *Proc. of 4th Conference, Faculty of Pure and Applied Sciences*. UWI, January. p 83.

[82] Sujad, N., Borana, K. and Manohar, S. (2014). "Effect of pesticide endosulfan on the growth of freshwater prawn, *Macrobrachium dayanum*" *International Journal of Pure and Applied Zoology*, *2*(3), 266-269.

[83] Wandan, E. N., Elleingand, E. F., Koffi, E., Clement, B. N. and Charles, B. (2010). "Impact of the insecticide endosulfan on growth of the African giant snail *Achatina achatina* (L.)". *African Journal of Environmental Science and Technology*, *4*(10), 685-690.

[84] Eriksen-Hamel, N. S. and Whalen, J. K. (2007). "Impacts of earthworms on soil nutrients and plant growth in soybean and maize agroecosystems". *Agriculture Ecosystems and Environment, 120,* 442–448.

[85] Dasgupta, R., Chakravorty, P. P. and Kaviraj, A. (2012). "Effects of carbaryl, chlorpyrifos and endosulfan on growth, reproduction and respiration of tropical epigeic earthworm, Perionyx excavatus (Perrier)". *Journal of Environmental Science and Health, Part B: Pesticides, Food Contaminants, and Agricultural Wastes, 47,* 2, 99-103.

[86] Van Beelen, P. and Doelman, P. (1997). "Significance and application of microbial toxicity tests in assessing ecotoxicological risks of contaminants in soil and sediment". *Chemosphere, 34,* 455–499.

[87] Rath, A. K., Ramakrishnan, B., Rath, A. K., Kumaraswamy, S., Bharathy, K., Singla, P. and Sethunathan, N. (1998). "Effect of pesticides on microbial biomass of flooded soil". *Chemosphere, 37*(4), 661–671.

[88] Amato, M. and Ladd, J. N. (1994). "Application of the ninhydrin-reactive N assay for microbial biomass in acid soils". *Soil Biology and Biochemistry, 26,* 1109-1115.

[89] Anderson, P. E. and Domsch, K. H. (1980). "Quantities of plant nutrients in the microbial biomass of selected soils". *Soil Science, 130,* 211-216.

[90] Quilchano, C. and Maranon, T. (2002). "Dehydrogenase activity in Mediterranean forest soils". *Biology and Fertility of Soils, 35,* 102-107.

[91] Kalyani, S. S., Sharma, J., Dureja, P., Singh, S. and Lata. (2010). "Influence of Endosulfan on Microbial Biomass and Soil Enzymatic Activities of a Tropical Alfisol". *Bulletin of Environmental Contamination and Toxicology, 84,* 351–356. Accessed November, 17 2018. DOI 10.1007/s00128-010-9943-x.

[92] Nare, R. W. A., Savadogo, P. W., Gnankambary, Z. and Sedogo, M. P. (2010). "Effect of Endosulfan, Deltamethrin and Profenophos on Soil Microbial Respiration Characteristics in Two Land Uses Systems in Burkina Faso". *Research Journal of Environmental Sciences*, 4, 261-270.

[93] Kirchner, M., Jakobi, G., Korner, W., Levy, L., Moche, W., Niedermoser, B., Schaub, M., Ries, L., Weiss, P., Antritter, F., Fischer, N, Henkelmann, B. and Schramm, K. (2016). "Ambient Air Levels of Organochlorine Pesticides at Three High Alpine Monitoring Stations: Trends and Dependencies on Geographical Origin". *Aerosol and Air Quality Research*, 16, 738–751.

[94] Estellano, V. H., Pozo, K., Harner, T., Franken, M. and Zaballa, M. (2008). "Altitudinal and Seasonal Variations of Persistent Organic Pollutants in the Bolivian Andes Mountains". *Environmental Science and Technology*, 42, 2528–2534.

[95] Li, Q., Wang, X., Song, J., Sui, H., Huang, L. and Li, L. (2012). "Seasonal and Diurnal Variation in Concentrations of Gaseous and Particulate Phase Endosulfan". *Atmospheric Environment*, 61, 620–626.

[96] Kuo, J. N., Soon, A. Y. m., Garrett, C., Wan, M. T. and Pasternak, J. P. (2012). "Agricultural pesticide residues of farm runoff in the Okanagan Valley, British Columbia, Canada". *Journal of Environmental Science and Health B*, 47(4), 250-61.doi:10.1080/03601234.2012.636588.

[97] Quinete, N., Castro, J., Fernandez, A., Zamora-Ley, I. M., Rand, G. M. and Gardinali, P. R. (2013). "Occurrence and distribution of endosulfan in water, sediment, and fish tissue: An ecological assessment of protected lands in South Florida". *Journal of Agriculture and Food Chemistry*, 61(49), 11881-11892. Doi;10.10211/jf403140z.

Chapter 6

STATUS OF ENDOSULFAN CONTAMINATION AND MANAGEMENT PRACTICES IN PAKISTAN

Mureed Kazim[1] and Jabir Hussain Syed[2],*
[1]Department of Environmental Sciences,
International Islamic University, Islamabad Pakistan
[2]Department of Meteorology, COMSATS University Islamabad Tarlai Kalan Park Road, 45550 Islamabad Pakistan

ABSTRACT

In the last two decades, many countries have recognized the hazards of persistent organic pollutants (POPs) and have banned or restricted their use and production. Among these persistent toxic substances, endosulfan has been widely used in the agriculture sector in Pakistan for more than 35 years to control pests. As a signatory of Stockholm Convention (SC), Pakistan has recently phased out; reduced/eliminated and destructed existing obsolete POPs pesticides across the country. This book chapter focuses on the current situation of endosulfan in Pakistan with the

---

* Corresponding Author's Email: jabir.syed@comsats.edu.pk.

emphasis on its historical usage and toxicity effects on environment and human. Needed policy and possible measures should be implemented at governmental level to avoid the increasing problem of endosulfan in the country. Available literature highlights that there is still a general lack of reliable data and research studies addressing endosulfan related issues in the context of environmental and human health risks in Pakistan. Therefore, there is a critical need to improve the current knowledge base, build upon the research experience from other countries which have experienced similar situations in the past. Further research into these issues in Pakistan is considered vital to help inform future policies/control strategies as already successfully implemented in other countries.

Keywords: endosulfan; historical usage, environmental risks, persistent organic pollutants, pests, human health

1. BACKGROUND

As the human population and food needs are increased with extreme pace during the last 2–3 decades, it exerts pressure on the agricultural system and environmental sanitation [1, 2]. Advanced crop production technology, an important component of the modern agricultural system has facilitated increasing population pressures. Use of agrochemicals has increased significantly in the recent past, which has resulted in environmental contamination [3]. Nowadays, it is unlikely to obtain high crop production without the use of pesticides, in addition to irrigation and chemical fertilizers. If the pesticides are not applied as per their recommended doses, the crop production would decrease significantly, and the food prices would soar dramatically. Under such circumstances, it would become impossible to sustain the whole human population. Therefore, the use of pesticides has become a vital component for increased crop production.

It has been estimated that 2.5 million tons of pesticides were being applied worldwide each year and increasing trends continue with the passage of time [4, 5]. Through linear risk extrapolation of animal data and maximum exposure levels of 550 million people, it was reported that there are 37,000 cancer cases yearly associated with pesticide use in developing countries [6]. Approximately three million people were poisoned and 200,000 died each year around the world from pesticide poisoning of which majority belongings to the developing countries [6, 7]. The dilemma is due to those pesticides that have been banned or recently being banned in many developed countries because of their toxic effects are still being used in the developing countries [8, 9]. It is also believed that in developing countries like Pakistan, the incidence of pesticide poisoning may even be greater than reported due to under-reporting and lack of data [10, 11]. Even in developed countries, despite the strict regulations and the use of safer pesticides, occupational exposures are significant [12].

Endosulfan is a pesticide belonging to the organo-chlorine group of pesticides, under the cycloidian subgroup. Introduced in the 1950s, it emerged as a leading agrochemical used against a broad spectrum of insects and mites in agriculture and allied sectors. It acts as contact and stomach poison and has a slight fumigant action [13]. It was used in vegetables, fruits, paddy, cotton, cashew, tea, coffee, tobacco and timber crops [14]. It was also used as a wood preservative and to control African flies and termites [15]. However, it is not recommended for household use. Intentional misuse of endosulfan for killing fish and snails has also been reported [16]. Endosulfan was also reported to be used deliberately as a method of removing unwanted fish from lakes before restoring. Endosulfan was introduced at a time when environmental awareness and knowledge about the environmental fate and toxicology of such chemicals were very low and not mandatory as per national laws [17].

Endosulfan is widely considered to be a Persistent Organic Pollutant (POP) but was not included in the initial list targeted for phase-out under the Stockholm Convention. Although the endosulfan was in the initial list of POPs being considered for worldwide elimination at the first meeting of experts in Vancouver, Canada (1994) jointly convened by governments of Canada and Philippines but was later removed from the list due to its moderate toxicity [18]. Later, endosulfan was listed as a POPs member in the Convention on Long-range Transboundary Air Pollution (LRTAP) 2007. Endosulfan is also recognized as a Persistent Toxic Substance (PTS) by the UNEP [19].

2. ENDOSULFAN: A GLOBAL ISSUE

Endosulfan has been in worldwide use since its introduction in the 1950s. It was considered as a safer alternative to other organochlorine pesticides in many countries since the 1970s. But in the last two decades, many countries have recognized the hazards of the wide application of this pesticide and have banned or restricted its use. Countries which have already banned endosulfan include Singapore, Belize, Tonga, Syria, Germany, Sweden, Philippines, Netherlands, St. Lucia, Columbia, Cambodia, Bahrain, Kuwait, Oman, Qatar, Saudi Arabia, United Arab Emirates, Sri Lanka and Pakistan [19]. However, restricted use was allowed in Australia, Bangladesh, Indonesia, Iran, Japan, Korea, Kazakhstan, Lithuania, Thailand, Taiwan, Denmark, Serbia & Montenegro, Norway, Finland, Russia, Venezuela, Dominican Republic, Honduras, Panama, Iceland, Canada, United States and United Kingdom ([19, 20].

It is one among the twenty-one priority compounds identified by the United Nations Environment Program (UNEP) and Global Environment Facility (GEF) in the Regional Based Assessment of Persistent Toxic Substances (PTS, 2002). These reports have considered the magnitude of use, environmental levels and human and ecological effects of these compounds [19]. Endosulfan has been banned in 8 nations of the Indian subcontinent. India is one of the major Indian Ocean rim nations, which has imposed no ban or restrictions on endosulfan. A ban on endosulfan exists in the South Indian state of Kerala (imposed through a Court Order), which came as an International POPs Elimination Network (IPEN) Pesticide Working Group Project-2004 Factsheet of public pressure following the poisoning of many villages due to aerial spraying of the chemical [21]. Colombia and Cambodia are two countries where endosulfan was banned for that time [20].

In pure form, endosulfan exists as colorless crystals. But the technical product is brownish crystals with a slight odor of Sulphur Dioxide [22]. Technically, endosulfan is a mixture of two isomers: i.e., α-endosulfan and β-endosulfan in the ratio 7:3. Technical grade endosulfan contains 94% - α-endosulfan and β-endosulfan and other related compounds like endosulfan alcohol, endosulfan ether, and endosulfan sulfate. Endosulfan is slightly soluble in water, but it dissolves readily in xylene, chloroform, kerosene, and most organic solvents and is a non-combustible solid. It is mixable with most fungicides and compatible with most pesticides [23].

The US Environmental Protection Agency (EPA) classifies endosulfan as Category Ib – Highly Hazardous. The European Union also rates it highly hazardous. World Health Organization (WHO) classifies endosulfan in Category II - Moderately Hazardous. The classification of WHO was found to be inappropriate considering the classification followed in many countries and the available toxicological information. It has been alleged that the classification is based mainly on LD50 value for acute toxicity generated by the producer company.

Worldwide use of endosulfan has been increased with the ban/restriction in use of the more persistent organo-chlorine pesticides like DDT and endrin. Endosulfan is acutely toxic and has been implicated in many cases of poisoning and fatalities. It has been identified with a range of chronic effects, including cancer and impacts on hormonal systems, exhibiting similarities with its predecessors in the organo-chlorine class.

3. ENDOSULFAN IN PAKISTAN

3.1. General Overview

Pakistan is primarily an agricultural country. The agriculture sector is considered as a lifeline of Pakistan's economy. According to the Economic Survey of Pakistan [24] this sector provides employment to > 42.3% of the population and is the major source of income for the majority of the country's inhabitants. Agriculture sector adds 19.5% of country's GDP [24]. Pesticides use is a common practice and an integrated component of crop production in Pakistan. According to Nation Master [25], the country is ranked 19th major utilizer of pesticide in the world. In Pakistan, pesticide uses are mostly focused on cotton-growing areas (80% of total pesticides used) [26, 27, 28]

Despite the above-mentioned data highlighting cotton production and pesticide use being important aspects of the country, the control of environmental pollution and associated health risks are not meticulous [29, 30, 31]. The low literacy rate of the country, especially of farmer community, is also contributing to the dilemma [32]. Like other developed countries such as European Union which has well-established regulations to deal with chemicals (such as REACH) [33], Pakistan does have legislations/regulations about pesticide use. However, the implementation stage of such legislations/regulations is poor in Pakistan [34].

Pakistan is one of the largest consumers of pesticides in the world after India and the 2nd highest among the South Asian countries. About 27% of the total pesticide consumption is used on fruit and vegetable crops [35]. In Pakistan, the annual consumption of endosulfan for vegetable crop production is higher than that of other pesticides. Pakistan became the signatory of Stockholm Convention on POPs on December 6, 2001, and then ratified it on April 14, 2008.

The key role and responsibility were assigned to the Ministry of Climate Change for compliance at the country level. According to the National Environmental Action Plan (NEAP) approved by the Pakistan Environmental Protection Council (PEPC) 2001, four areas which are highly sensitive in environmental context need immediate attention. These include fresh air, clean water, treatment of solid wastes and ecosystem management. To observe the compliance of obligations set under the Stockholm Convention, 1st National Implementation Plan (NIP) of Pakistan was executed in 2006-2009. The second NIP is under preparation from 2015-2019 under the Ministry of Climate Change. Europe has dedicated regulation on POPs that can be assessed as an example of standard rules and regulations to address growing threats of POPs in Pakistan.

In many developing countries including Pakistan, a number of malpractices in the agricultural system result in severe environmental contamination [36, 37]. Some of the common malpractices adopted by the Pakistani farmers include spraying of pesticides with limited prior knowledge and ignoring basic safety measures; trigger several environmental concerns in the country [38, 39].

Past studies have shown substantial cases of misuse or overuse of pesticides by the farming community in Pakistan, especially in the cotton-growing belt of the country [40]. For example, some farmers in Punjab were using 70% more pesticide compared to their counterpart to obtain same cotton production [41]. Similarly, in Sindh province, a great number of farmers were not using pesticide at the appropriate time of spray. Recently, [42] reported that 49% of the farmers in two districts of Punjab were overusing pesticides than they required. Some studies also reported the lack of knowledge of farmers about pesticide use in Pakistan [42]. Consequently, pesticide use is not properly implemented in the country due to ineffective legislation, lack of awareness, and technical know-how/training among the people in general and farming community.

Table 1. Top Ten Insecticides in Terms of Weight

Insecticide	Sale of pesticides in weight (MT)	%
Methamidophos	277.4	28.6
Endosulfan	117.98	12.2
Cypermetherin	90.92	9.4
Imidacloprid	73.5	7.6
Fenpropthrin	63.98	6.6
Chlorpyrifos	44.93	4.6
Bifenthrin	41.7	4.3
Profenophos	36.7	3.8
Fenvalerate	27.6	2.8
Monocrotophos	22.3	2.3
Others	173.74	17.9
Total	970.75	100

3.2. Endosulfan Usage

Endosulfan has been one of the top ten pesticides with most sales in the Pakistani market in terms of weight during 2003 (Table 1) [43].

These estimates support the survey findings of the study conducted by [44] who reported that most of the pesticides applied in Pakistan were category I as classified by the WHO like monocrotophos, methamidophos, endosulfan, and carbofuran.

As an insecticide, endosulfan has been used only in agriculture in Pakistan for more than 35 years to control pests. In 1980, the pesticide including endosulfan business was transferred to the private sector with the agreement that the pesticides available in government stock will not be imported until they are exhausted. The pesticide consumption figures for 1992 stand at 5519 metric tons active ingredient but recent consumption reports are missing.

Insecticides comprise 85% of the total pesticides and the cotton crop is the major recipient of these chemicals. Endosulfan became a highly controversial agrichemical due to its acute toxicity, potential for bioaccumulation, and role as an endocrine disruptor. Because of its threats to human health and the environment, a global ban on the manufacture and use of endosulfan was negotiated under the Stockholm Convention in April 2011. The ban has taken effect in mid-2012, with certain uses exempted for five additional years. More than 80 countries had already banned it or announced phase-outs by the time the Stockholm Convention ban was agreed upon. It is still being used extensively in India, China, Pakistan, and a few other countries.

Endosulfan has been used in the agriculture sector of Pakistan to control insect pests including whiteflies, aphids, leafhoppers, mainly cotton pests. Due to its unique mode of action, it is useful in resistance management; however, as it is not specific, it can negatively impact populations of beneficial insects. It is, however, considered to be moderately toxic to honey bees and it is less toxic to bees than organophosphate insecticides.

3.3. Contamination Status

Due to the lack of any statistical data about the contamination levels of Endosulfan in Pakistan, it is difficult to endorse the range of toxicity that is reported globally, however, some case studies have been conducted that are evidence of its toxic effects in Pakistan (Table 2). Endosulfan is acutely neurotoxic to both insects and mammals, including humans. EPA's acute reference dose for dietary exposure to endosulfan is 0.015 mg/kg for adults and 0.0015 mg/kg for children. For chronic dietary exposure, the EPA references doses are 0.006 mg/(kg·day) and 0.0006 mg/(kg·day) for adults and children, respectively.

3.3.1. Endosulfan in Biological Samples

Toxic effect of endosulfan on aquatic life has been reported globally as well as in Pakistan. Fish ecosystem suffer more like Labeo rohita is considerably sensitive to endosulfan by exhibiting significantly higher genetic damage index (GDI), percent damaged the cell and a cumulative tail length of the Comets. The length of cumulative tail even at the lowest sub lethal concentration demonstrated higher susceptibility of rohu to endosulfan. Impact of endosulfan shows the inability of rohu to interact against Endosulfan and potential danger to carp's survival in natural water bodies [45].

The levels of DDT, DDE, endosulfan, endosulfan sulfate, carbofuran, and cartap which were estimated in the flesh of Catla catla fish sampled from ten sites of Ravi River between its stretches from Shahdara to Head Balloki Punjab, Pakistan was reported positively. Mean annual minimum endosulfan concentration in Catla catla was detected as $0.112 + 0.003 \mu g\, g - 1$ at the Shahdara bridge, whereas maximum mean annual concentrations of endosulfan ($0.136 + 0.0031\, \mu g\, g - 1$) was detected at after Degh fall river sampling site, respectively. The difference between these two river sampling sites, for the toxicity of endosulfan, was statistically highly significant ($P < 0.01$).

The lowest mean concentration of $0.099 + 0.002\, \mu g\, g - 1$ of endosulfan was detected in Catla catla during the month of August 2010, while the highest mean concentration of $0.0137 + 0.002\, \mu g\, g - 1$ of endosulfan in Catla catla was detected during the month of January 2010. The difference between these two months, for the toxicity level of endosulfan, was statistically highly significant ($P < 0.01$) [46].

Table 2. Summary of contamination levels of Endosulfan in Pakistan

S. NO.	Year of Study	Study area	Matrices	Studied Pesticides	Major Findings	References
1	2003	Karachi	Top 10 pesticides sales in Karachi	Ten pesticides including Endosulfan	12.2% endosulfan	Khooharo et al. (2008)
3	2018	Vehari District, Punjab	Soil	α- and β-endosulfan	up to 14 µg/mg	Ahmed et al. (2018)
4	2008	Swat Valley, KP Province	Water	Endosulfan and Cypermethrin	0 to 12 mg/kg endosulfan	Nafees et al. (2008)
5	2004	Bahawalpur, Rajanpur, Muzaffargarh and Dera Ghazi Khan Areas	Water/Soil	Bifenthrin, E-cyhalothrin, methyl parathion and endosulfan	Endosulfan reported 30% of the total samples	Tariq et al. (2004)
6	2017	In Vivo Study	Fish (Labeo rohita)	Endosulfan	Significant higher genetic damage reported	Ullah et al. (2017)
7	2014	Ravi River Punjab	Catla Catla Fish	DDT, DDE, endosulfan, and carbofuran	Concentration of endosulfan 0.112 to 0.136µg g−1	Akhtar et al. (2014)
8	2010/2011	Hyderabad region, Sindh	Apple, Grape, Orange	Chlorpyrifos, dieldrin, endosulfan sulfate, parathion, Disulfoton and Tiradimefo	Higher concentration of endosulfan (1,236 ng/g) in apple	Latif et al. (2011)
9	2010	Nawab Shah, Sindh	Eight Fruit Samples	Endosulfan	Only apple samples were exceeding MRL	Latif et al. (2011)
10	2006	Faisalabad	Vegetables	Endosulfan	Peeling reduce ER in potato 67%, brinjal (60%).	Randhawa at al. (2007)
11	2016	Faisalabad, Gujranwala and Multan, Punjab	Okra and Brinjal	Endosulfan	60% okra, 46.66% brinjal were contaminated	Randhawa et al. (2016)
12	1986	Quaid-I-Azam University Islamabad, Pakistan	Cyprinid Fish (Cyprinion watsoni)	Endosulfan	Harmful effects on histology of the testes	Kalsoom et al. (2005)

S. NO.	Year of Study	Study area	Matrices	Studied Pesticides	Major Findings	References
13	2009	Lab Study	Peach nut shells	Endosulfan	Methanol was better solvent to adsorbed endosulfan	Memon et al. (2009)
15	2007	14 Districts of Sindh, Pakistan	Agriculture spray-workers	Endosulfan, monocrotophos, carbaryl, and cypermethrin	(44.28%) workers suffer from averagely 0.009 mg/kg	Soomro et al. (2008)
17	1998 - 2007	Cotton growing areas of Pakistan	Beet army worm, Spodoptera exigua	Endosulfan, chlorpyrifos, quinalphos, pyrethroids cypermethrin, deltamethrin, bifenthrin, and fenpropathrin	Endosulfan resistance declined from 2000 to 2006	Ahmad et al. (2010)
18	1996-2004	Central Cotton Research Institute, Multan	Susceptibility of Aphis gossypii	Endosulfan, Organophosphates, and Carbamates	No resistance was found against endosulfan	Ahmed and Arif (2008)
19	2006-2010	Nuclear Institute for Agriculture and Biology, Faisalabad, Pakistan	Susceptibility of aphid Brevicoryne brassicae	Endosulfan, Organophosphates, and Neonicotinoids	No or very low levels of resistance was found	Ahmad and Akhtar (2013)
20	2009	Faisalabad, Punjab	Meat and Organs of cattle	Chlorpyrifos, Cyhalothrin, Cypermethrin, and Endosulfan	2% of samples contaminated with endosulfan	Muhammad et al. (2010)
21	2014		Salmonella strains	Endosulfan and Lambda-cyhalothrin	Increase in the mutagenicity was detected in one strain	Saleem et al. (2014)
22	1998	Rice-growing area in Punjab Province	Impact of neem cake in soil against endosulfan	Endosulfan	Prolonged the period of degradation	Akhter et al. (1998)

Kalsoom et al. (2005) assessed the effects of endosulfan, an organochlorine insecticide, on reproductive and developmental parameters of male fresh water Cyprinid fish, Cyprinion watsoni from Ramli stream in hilly areas near Quaid-I-Azam University Islamabad, Pakistan during early spawning period (March). Low dose (0.75 ppb) treatment of endosulfan had a positive effect on the increase in testicular length, breadth, and weight, while treatment with high dose (1 ppb) endosulfan had no significant effects on these parameters. This suggests that high dose of endosulfan (> 0.75 ppb) does not bring in major changes in morphology of the testes, but it does have more harmful effects on histology of the testes.

Soomro et al. (2008) studied the agriculture farmers those were professionally spray-workers from 14 districts; Badin, Dadu, Ghotki, Hyderabad, Jacobabad, Khairpur, Larkana, Mirpurkhas, Nawabshah, Nosheroferoz, Sanghar, Shikarpur, Sukkur, and Thatta in Sindh province of Pakistan. Areas were selected according to the repeated sprays of pesticides; carbaryl, cypermethrin, Endosulfan, and monopcrotophos on crops of cotton, tomato, chili, okra vegetable and mango archerds during the year 2005 harvest season. As endosulfan, a product of organo-chlorine pesticides is lipophilic in nature, its presence in human tissues and fluids is reported to cause neurotoxic effects. Its bioavailability may produce toxic effects, as the highest percentage of spray-workers (44.28%) having Endosulfan residues averagely 0.009 mg/kg bodyweight, so this contamination of blood can be predicted for environmental effects on their health.

Ahmad et al. (2008) assessed for their resistance effects of chlorinated hydrocarbon endosulfan, the organophosphates chlorpyrifos, and quinalphos, and the pyrethroids cypermethrin, deltamethrin, bifenthrin and fenpropathrin in beet armyworm i.e., Spodoptera exigua (Lepidoptera: Noctuidae), from Pakistan. Using a leaf-dip bioassay, resistance to endosulfan was high during 1998-2000 but declined to very low, too low levels during 2001-2007, following a reduced use of the insecticide. S. exigua is an early-season pest of cotton in Pakistan. Endosulfan was used to be applied early in the season against sucking insect pests, and therefore S. exigua was frequently exposed to endosulfan. A moderate to high resistance was found to endosulfan in five field populations of S. exigua tested during

1998-2000. The Shershah-1 population rather exhibited very high resistance to endosulfan. Endosulfan resistance declined to low levels during 2000-2002 and then to very low levels during 2003-2006. This drop in endosulfan resistance may be associated with a decreased use of endosulfan against insect pests of cotton in Pakistan.

Twenty-six pesticides were reported in three popular fruits (apple, grapes, orange) collected from local fruit markets of Hyderabad region, situated in the province of Sindh, Pakistan. Out of the total 131 analyzed samples, 53 (40%) were found containing pesticide residues while only three (2%) samples exceeded the maximum residue limits (MRLs) of some pesticides. Endosulfan sulfate (1236 ng g-1), Chlorpyriphos (1,256 ng g-1) were found higher in apple samples [50]. Similarly, contamination of fruits samples of the local markets of Nawab Shah, District Sindh Pakistan show contamination of endosulfan along with organophosphates, pyrethroid, and organochlorines. Eight fruit samples of apple, guava, orange, grapes, pear, persimmon, banana, and pear purchased [51]. All the fruit samples were found contaminated except banana, and among this only apple, samples were found exceeding the MRLs (2000 µg kg-1 set by Codex Alimentarius Commission [52].

Washing, peeling and heat treatment to a certain vegetable resulted in the removal of endosulfan residues (ER). The results show that ER removal was 67% and 60% of in potato and brinjal, respectively [53]. Heat treatment during cooking had a significant effect on the reduction (13-35%) of ER in all the tested vegetable samples. The effect was more pronounced in okra and tomato while the least effect was observed in potato and brinjal, which may be due to the nature of different matrices. However, present results agree fairly with previous findings of [54, 55, 56].

Exposure of organochlorine pesticides to the urban population of Faisalabad, Gujranwala, and Multan, Punjab, Pakistan was tested as a result of the utilization of okra and brinjal. A total of 180 samples of vegetables were collected and samples were analyzed for pesticide residues i.e., α-endosulfan, β-endosulfan, endosulfan sulfate, HCH, Y-HCH, DDE and DDT. 30 samples of okra and brinjal were analyzed in triplicate for each pesticide residue i.e., α-endosulfan, β-endosulfan, endosulfan sulphate,

HCH, gamma HCH, DDE and DDT. Mean residue levels for detectable α-endosulfan in okra and brinjal were as follow: Faisalabad, 0.0229 mg kg^{-1} and 0.0172 mg kg^{-1}, Gujranwala, 0.1122 mg kg^{-1} and 0.1008 mg kg-1 Multan, 0.0625 mg kg^{-1} and 0.0534 mg kg^{-1}. Overall 57, 50 and 60% samples of okra, and 47, 40 and 47% samples of brinjal were found contaminated with α-endosulfan [57].

Market-based survey was carried out to evaluate the level of 26 pesticides in some commonly used fruits in Hyderabad region, Pakistan [58]. Gas chromatography coupled with micro electron capture detector was used to assess the levels of pesticide residues. In this study, the residues of targeted pesticides were evaluated in 131 samples of apple, grapes and orange obtained from the three fruit markets i.e., towns Latifabad (market number 1), Qasimabad (market number 2) and main Hyderabad city (market number 3). In the analyzed samples, 7 pesticides belonging to the different chemical groups (organophosphates, organochlorines, and triazole) with different properties (6 insecticides and 1 fungicide) were detected. The total number of samples collected from each market, identified classes of pesticides and numbers of samples above to the MRLs are illustrated in Table 6. Out of the total 131 samples analyzed, 53 samples (40%). The outcomes of the present study authenticated the existence of pesticides such as chlorpyrifos, dieldrin, endosulfan sulfate, parathion, disulfoton, and triadimefon in fruit samples which were applied in pre-harvest treatment. To avoid adverse effects on public health, it is a necessity to set up control measures to make sure that each pesticide should be below MRLs in the fruits to be marketed. The study presented significant information regarding pesticide residues contamination on fruits from the Hyderabad region.

Latif et al. (2011) determined the contamination of pesticides in the meat and organs of cattle reared in pesticide spraying areas of Faisalabad, Pakistan. Because no such published information is available in this region about such case. The meat and organs such as liver, lung, and kidney were collected from villages situated within the radius of 25-35 km on four different localities (Pensara, Aminpur, Jaranwala, and Sheikhupura roads) in the Northeast and Southwest of the city during winter and spring seasons of 2009. Five pesticides (cyhalothrin, endosulfan, chlorpyrifos, cypermethrin, and methyl parathion) were analyzed in the collected meat and organs (n = 600) with solid phase microextraction and high-performance liquid chromatography techniques. The residue analysis revealed that about 13, 21, 4, and 2% muscle samples were contaminated with chlorpyrifos, cyhalothrin, cypermethrin, and endosulfan, respectively.

3.3.2. Endosulfan in Non-Biological Samples

The concentration of residual pesticides rises in Lower Swat, Pakistan as thousands of kg of pesticides are applied in this upper part of Swat Valley and hence are washed into the Swat River, which is used for irrigation downstream. Dichlorvos increased from 47mg/kg in the north to 159 mg/kg in the south, endosulfan from 0 to 12 mg/kg, Methidiathion from 38 to 125 mg/kg, and Cypermethrin from 43 to 184 mg/kg [59]. Irrigated water, after passing agriculture fields with such residual pesticides concentration (RPC), appeared as a potential hazard when joining the mainstream of the Swat river in Mardan district, situated downstream of the Swat River, KPK Province, Pakistan. The most widely used pesticides identified were Cypermethrin (26 – 184 mg/kg) and Endosulfan (6.24 – 12.6 mg/kg) [59]. Also, contamination of endosulfan and some other pesticides in the surface and ground water from different agricultural districts of Pakistan Viz; Bhawalnagar, Rajanpur, Muzafarghar, and Dera Ghazi Khan were positively observed [60]. The surface/ground water via surface run-off from agricultural fields, direct spray on water ponds to control mosquitoes and other insect to protect animals, infiltration of different chemicals via heavily rainfall endosulfan was reported which occurred in more than 30% of the total samples and ranged from 0.0 to 0.02 g/L [60]. The use of high doses of pesticides in tobacco fields (8 – 10 sprays on each crop), which may reach the groundwater by infiltration of these chemicals into deep soil due to heavy rainfall [61, 60].

A recent case study was conducted to obtain a concrete figure about the presence of endosulfan in Vehari district, Punjab, Pakistan. The results showed the high levels of endosulfan in the soil (up to 14 μg/mg). High levels of α- and β- endosulfan in the soil can be due to low mobility of α- and β- endosulfan in the soil that supports the previous studies of high levels of endosulfan in soils and sediments [62, 63]. Adsorption of α- and β-endosulfan in the upper soil layer by organic and inorganic soil components reduce their mobility and percolation through the soil profile. Endosulfan is a low water soluble and polar pesticide, which is absorbed hydrophobically to the soil [64] and may persist for a long time in the soil as bound residues particularly associated with soil organic matter [65, 66].

3.4. Toxicity Assessment

Pakistani field populations of Aphis gossypii were assessed from 1996 to 2004 for their susceptibility to endosulfan, organophosphates (monocrotophos, dimethoate, profenofos, chlorpyrifos, quinalphos, parathion-methyl, pirimiphos-methyl, and ethion) and carbamates (carbaryl, methomyl, thiodicarb, furathiocarb, and carbosulfan) using a leaf-dip bioassay method. Generally, there was a very low resistance to endosulfan, monocrotophos, profenofos, chlorpyrifos, quinalphos, pirimiphos-methyl, carbaryl and methomyl, and a low to moderate resistance to dimethoate, parathion-methyl, and thiodicarb. Some of the populations had a very high resistance to parathion-methyl, ethion, and thiodicarb. However, no resistance was found to the carbamate aphidicides furathiocarb and carbosulfan. Correlation analysis demonstrated the positive correlation of LC50s within but not between the two insecticide groups (1) endosulfan, profenofos, chlorpyrifos and parathion-methyl and (2) monocrotophos, dimethoate, pirimiphos-methyl, ethion, carbaryl, methomyl and thiodicarb [67].

Cabbage aphid Brevicoryne brassicae (L.) (Hemiptera: Aphididae) is a serious pest of crucifers in Pakistan. After incidences of poor control by recommended insecticides, the study was undertaken to find out the status of insecticide resistance in Pakistani B. brassicae [68]. Apterous adult aphids were bioassayed from 2006 to 2010 for their response to 12 insecticides using an adult immersion method. No or very low levels of resistance was found to endosulfan; and the organophosphates: chlorpyrifos and profenofos. Resistance to methomyl; emamectin benzoate; the pyrethroids:

cypermethrin, lambdacyhalothrin, bifenthrin and deltamethrin; and the neonicotinoids: imidacloprid, acetamiprid, and thiamethoxam; increased progressively in concurrence with their regular use on vegetables. B. brassicae resistance to these insecticides remained very low to low in 2007 and 2008, but then it increased to moderate to high levels in 2009 (except cypermethrin and bifenthrin) and 2010. Under heavy infestations of this aphid, the application of insecticides having no, very low and low resistance is recommended in the rotation.

Ahmad et al. (2013) investigated the genotoxic and cytotoxic potential of Endosulfan (EN) and Lambda-cyhalothrin (LC); individually and in combination. 3-(4,5-Dimethylthiazol-2-yl)-2,5-Diphenyltetrazolium Bromide (MTT) assay was utilized to determine cytotoxicity, while two mutant histidine dependent Salmonella strains (TA98, TA100) were used to determine the mutagenicity of EN and LC. Moreover, mutagenicity assay was conducted with and without S9 to evaluate the effects of metabolic activation on mutagenicity. Even though a dose dependent increase in the number of revertant colonies was detected with EN against both bacterial strains, a highly significant ($p < 0.05$) increase in the mutagenicity was detected in TA98 with S9.

Saleem et al. (2014) studied the influence of two types of neem cake (solvent-extracted, NC-I and expeller-extracted, NC-II) on the persistence in soil of endosulfan and diazinon applied as commercial formulations. It was found that both types of neem cake applied at 10, 20 or 30 g ha-1 prolonged the period of degradation as compared with soils without neem cake amendment, and hence increased the persistence of the insecticides. There was little difference in the effect of the two types of neem cake. Treatment of the soil with insecticide 10 days after amendment with neem cake did not lead to any increase in persistence.

Akhtar et al. (1998) studied the adsorption efficiency of peach nut shells for the removal of endosulfan from aqueous solutions and suggested application of peach nut shells as a cheap and abundant agricultural by-product in Pakistan for the removal of endosulfan from the water. The adsorption of endosulfan has also been studied as a function of contact time, concentration and temperature. Low-cost peach nut shells were found to be effective for the removal of endosulfan pesticide from aqueous solutions. The value of 1/n from Freundlich adsorption isotherm proposes adsorption capacity of peach nut shells is better for lower concentration solutions rather than higher concentration solution. The thermodynamic quantities ΔH and ΔG indicate the endothermic and spontaneous nature of the adsorption process. Endosulfan was successively removed from surface water samples. Methanol was found to be better solvent to desorb the adsorbed endosulfan from the surface of peach nut shells.

3.5. Remediation of Endosulfan

Biodegradation of endosulfan by certain microorganism species such as Aspergillus niger isolated from cotton fields of Punjab, Pakistan indicates the potential to degrade endosulfan effectively within a short period of time. It can tolerate high concentrations of endosulfan but up to a certain limit and can completely degrade 0.1% endosulfan in four days. Moreover, the study indicates that the inoculum size had no significant effect while the pH of the media had a profound effect on the degradation of endosulfan [72]. Similarly, biodegradation of endosulfan was tested through bacterial strain Stenotrophomonas maltophilia EN- 1 that was isolated from the agriculture-contaminated soil of Shujaabad, Multan, Pakistan. Degradation of both isomers i.e., α– and β– endosulfan and approximately 70% of the initial concentrations was degraded within 5 days of incubation. The result also showed that the higher concentrations irrespective of isomers take more time to degrade. The rate decreased with the addition of supplementary sulfur source. Therefore, it clearly indicates that S. maltophilia is highly efficient to utilize endosulfan as the sole source of sulfur [73].

Biodegradation of α– and β–endosulfan in soil slurries from the bacterial strain of Pseudomonas aeruginosa was tested. Endosulfan degradation was most effectively achieved at an initial inoculum size of 600 μl (OD = 0·86), incubation temperature of 30°C, in aerated slurries at pH 8, in loam soil. Under these conditions, the bacterium removed more than 85% of spiked α– and β– endosulfan (100 mg l–1) after 16 days. Abiotic degradation in non-inoculated control medium within the same incubation period was about 16%. Biodegradation of endosulfan varied in different textured soils, being more rapid in coarse textured soil than in fine textured soil. Increasing the soil contents in the slurry above 15% resulted in less biodegradation of endosulfan. Exogenous application of organic acids (citric acid and acetic acid) and amino acids had stimulatory and inhibitory effects, respectively, on biodegradation of endosulfan [73].

Four strains of Helicoverpa armigera showed a very low level of resistance in populations against endosulfan as compared from Bahawalpur and Multan and moderate resistance in Dera Ghazi Khan and Rahim Yar Khan region. Out of the four strains tested, Bahawalpur strain exhibited very low resistance. Multan and D. G. Khan strains exhibited a low resistance, but Rahim Yar Khan strain manifested a moderate resistance as compared to the Lab-PK [74].

Intensive use of endosulfan has resulted in contamination of soil and water environments at various sites in Pakistan. A study was conducted to isolate efficient endosulfan-degrading fungal strains from contaminated soils. Sixteen fungal strains were isolated from fifteen specific sites by employing enrichment techniques while using endosulfan as a sole sulfur source and tested for their potential to degrade endosulfan. Biodegradation of endosulfan by soil fungi was accompanied by a substantial decrease in pH of the broth from 7.0 to 3.2. The major metabolic product was endosulfan diol along with very low concentrations of endosulfan ether. Maximum biodegradation of endosulfan by these selected fungal strains was found at an initial broth pH of 6, incubation temperature of 30°C and under agitation conditions.

Arshad et al. (2008) studied the sorption behavior of silica resin to remove the endosulfan isomers from aqueous solution. The efficiency of resin was checked through both batch and column sorption methods. Sorption efficiency of immobilized silica resin was investigated to remove endosulfan isomers from aqueous solution. Results of the sorption experiment show that silica resin is more efficient than pure silica. The sorption process was found to be pH dependent. Other factors, like, contact time and shaking speed also affect sorption efficiency of synthesized resin.

Sorption results showed that Freunlich sorption isotherm was best fitted to sorption data following the pseudo-second-order rate equation. On the other hand, Thomas model suggests that column sorption is more efficient than batch method. An excellent regeneration for at least five times with slight up-down in sorption capacity indicates that resin is highly applicable for the removal of endosulfan from agriculture/industrial waste water.

4. Summary and Conclusion

This chapter summarized and discussed the current status of endosulfan usage, its contamination levels, toxicity assessment, and fresh management practices to remediate the endosulfan issue in Pakistan. Although the available published data is limited but based on the research studies, it can be concluded that this issue has been threatening to both human and environmental health for the last many years. Extensive application of pesticides/insecticides has seriously affected food quality. Being a signatory of the Stockholm Convention on persistent organic pollutants (POPs), it has been observed that efforts have been made to phase out and eliminate persistent toxic substances including endosulfan from Pakistan. In addition, remediation practices are also being used to get rid of endosulfan' traces from the environment. It has been also noted that though alternatives to endosulfan are available, application of the alternative non-chemical pest control in Pakistan and other parts of the world is the need of the hour. If governments and research institutions can support such work, the use of endosulfan can be eliminated in agriculture and other sectors. Adopt "Land and Food without Poisons" as a goal for survival and adopt organic, ecological or natural agriculture. To sum up, this chapter reveals that it is not yet prepared to face and manage the challenges ahead, attributed to the increasing use of pesticides/insecticides and their potentially harmful effects on environmental and human health. It is, therefore, a critical need for further research into these issues in Pakistan to obtain a more holistic and nuanced perspective on relevant issues and to further compare the situation in Pakistan versus other countries.

REFERENCES

[1] Mombo, S., Dumat, C., Shahid, M. & Schreck, E. (2017). A socio-scientific analysis of the environmental and health benefits as well as potential risks of cassava production and consumption. *Environ Sci Pollut Res*, *24*, 5207–5221.

[2] Natasha, S. M., Niazi, N. K., Khalid, S., Murtaza, B., Bibi, I. & Rashid, M. I. (2018). A critical review of selenium biogeochemical behavior in soil-plant system with an inference to human health. *Environ Pollut*, *234*, 915–934.

[3] Xiong, T., Dumat, C., Dappe, V., Vezin, H., Schreck, E., Shahid, M., Pierart, A. & Sobanska, S. (2017). Copper oxide nanoparticle foliar uptake, phytotoxicity, and consequences for sustainable urban agriculture. *Environ Sci Technol*, *51*, 5242–5251.

[4] Pimentel, D. (1995). Amounts of pesticides reaching target pests: environmental impacts and ethics. *J Agric Environ Ethic*, *8*, 17–29.

[5] Food and Agricultural Organization (FAO). FAO/WHO global forum of food safety regulators. Marrakech, Morocco, 28–30 January; 2002. [http://www.fao.org/DOCREP/MEETING/004/AB428E.HTM Agenda Item 4.2 a, GF/ CRD Iran-1].

[6] WHO/UNEP Working Group. Public health impact of pesticides used in agriculture. Geneva: World Health Organization, 1990.

[7] Food and Agricultural Organization (FAO). Project concept paper. HEAL: health in ecological agricultural learning. Prepared by the FAO programme for community IPM in Asia. Rome: Food and Agricultural Organization of the United Nations; 2000. [http://www.fao.org/nars/partners/2nrm/proposal/ 9-2-6.doc.].

[8] Wilson, C. & Tisdell, C. (2001). Why farmers continue to use pesticides despite environmental, health and sustainability costs. *Ecol Econ*, *39*, 449–62.

[9] Sankararamakrishnan, N., Sharma, A. K. & Sanghi, R. (2005). Organochlorine and organophosphorous pesticide residues in ground water and surface waters of Kanpur, Uttar Pradesh, India. *Environ Int*, *31*, 113–20.

[10] Forget, G. (1991). Pesticides and the third world. *J Toxicol Environ Health*, *32*, 11–31.

[11] Tariq, M. I. (2005). *Leaching and degradation of cotton pesticides on different soil series of cotton growing areas of Punjab, Pakistan in Lysimeters*. Ph. D. Thesis, University of the Punjab, Lahore, Pakistan.

[12] World Resources Institute (WRI). (1998). *World resources, 1998/1999*. UK: Oxford University Press.

[13] Nayar, K. K., Ananta Krishnan, T. N. & David, B. V. (1989). *General and Applied Entomology*. Tata McGrow Hill Publishing Co Ltd, New Delhi, India., 17a.

[14] Michael, Hermann. (2003). *Endosulfan Preliminary Dossier*; www.unece.org/env/popsxg/docs/2000-2003/dossier-endosulfan may03.pdf.

[15] Anon. (1984). Environment Health Criteria 40- Endosulfan. IPCS (International Programme on Chemical Safety) – WHO Geneva.

[16] Anon. (2002). *End of the Road for Endosulfan- A Call for Action Against A Dangerous Pesticide Environmental Justice Foundation*, London, UK. Internet site- www.ejfoundation.org.

[17] *PANUPS Endosulfan Residue in Australian Beef (Agro: World Crop Protection News*, Aug 28- 1998, Jan 15 1998), PANUPS (Pesticide Action Network- Update Service- May 20 1996) 17 g.

[18] Romeo, F. & Quijano, M. D. (Oct/Dec 2000). Risk Assessment in a third world reality: An Endosulfan case History. *International Journal of Occupational and Environment Health.*, Vol. 6, No. 4, 17 h.

[19] Anon. (Dec 2002). *Regional Based Assessment of Persistent Toxic Substances- South East Asia and South Pacific- Regional Report – Chemicals-* United Nations Environmental Programme- Global Environment Facility.

[20] Anon. (June 2003). Pesticide News No 60, *The Journal of Pesticide Action Network UK*. (Quarterly), p. 19.

[21] De Munari, A., Semiao, A. J. C. & Antizar-Ladislao, B. (2013). Retention of pesticide endosulfan by nanofiltration: influence of organic matter- Environ Sci Pollut Res pesticide complexation and solute-membrane interactions. *Water Res*, 47, 3484–3496.

[22] Muhammad, F., Akhtar, M., Rahman, Z. U., Farooq, H. U., Khaliq, T. & Anwar, M. I. (2010). Multi-residue determination of pesticides in the meat of cattle in Faisalabad-Pakistan. *Egypt. Acad. J. Biol. Sci*, 2, 19-28.

[23] India (Oct 2001) Preliminary findings of the survey on the impact of aerial spraying on the people and the ecosystem: Long Term Monitoring – *The Impact of Pesticides on the People and Ecosystem (LMIPPE)*.

[24] Chattopadhyay, S. B. (1993). *Principles and Procedures of Plant Protection* (Oxford and IBH Publishing Co. Pvt Ltd, New Delhi), 85-86.

[25] Master, N. (2016). Pesticide use: Countries Compared.

[26] Khan, M., Akram, N., ul Husnain, M. I. & Qureshi, S. A. (2011). Poverty environment nexus: use of pesticide in cotton zone of Punjab, Pakistan. *J Sustain Dev*, 4, 163.

[27] Rehman, A., Jingdong, L., Chandio, A. A., Hussain, I., Wagan, S. A. & Memon, Q. U. A. (2016). Economic perspectives of cotton crop in Pakistan: a time series analysis (1970–2015)(part 1). *J Saudi Soc Agric Sci*.

[28] Shahid, M., Ahmad, A., Khalid, S., Siddique, H. F., Saeed, M. F., Ashraf, M. R., Sabir, M., Niazi, N. K., Bilal, M. & Naqvi, S. T. A. (2016). *Pesticides pollution in agricultural soils of Pakistan, soil science: agricultural and environmental prospectives*. Springer, pp. 199–229.

[29] Baqar, M., Sadef, Y., Ahmad, S. R., Mahmood, A., Li, J. & Zhang, G. (2018). Organochlorine pesticides across the tributaries of river Ravi, Pakistan: human health risk assessment through dermal exposure, ecological risks, source fingerprints, and spatiotemporal distribution. *Sci Total Environ*, 618, 291–305.

[30] Saeed, M. F., Shaheen, M., Ahmad, I., Zakir, A., Nadeem, M., Chishti, A. A., Shahid, M., Bakhsh, K. & Damalas, C. A. (2017). Pesticide exposure in the local community of Vehari District in Pakistan: an assessment of knowledge and residues in human blood. *Sci Total Environ*, 587, 137–144.

[31] Waheed, S., Halsall, C., Sweetman, A. J., Jones, K. C. & Malik, R. N. (2017). Pesticides contaminated dust exposure, risk diagnosis and exposure markers in occupational and residential settings of Lahore, Pakistan. *Environ Toxicol Pharmacol*, 56, 375–382.

[32] Saeed, M. F., Shaheen, M., Ahmad, I., Zakir, A., Nadeem, M., Chishti, A. A., Shahid, M., Bakhsh, K. & Damalas, C. A. (2017). Pesticide exposure in the local community of Vehari District in Pakistan: an assessment of knowledge and residues in human blood. *Sci Total Environ*, 587, 137–144.

[33] Shahid, M., Pinelli, E. & Dumat, C. (2018b). Tracing trends in plant physiology and biochemistry: need of databases from genetic to kingdom level. *Plant Physiol Biochem*, 127, 630–635.

[34] Shahid, M., Ahmad, A., Khalid, S., Siddique, H. F., Saeed, M. F., Ashraf, M. R., Sabir, M., Niazi, N. K., Bilal, M. & Naqvi, S. T. A. (2016). *Pesticides pollution in agricultural soils of Pakistan, soil science: agricultural and environmental prospectives*. Springer, pp. 199–229.

[35] Damalas, C. A. & Khan, M. (2016). Farmers' attitudes towards pesticide labels: implications for personal and environmental safety. *Int J PestManag*, 62, 319–325.

[36] Khalid, S., Shahid, M., Dumat, C., Niazi, N. K., Bibi, I., Gul Bakhat, H. F. S., Abbas, G., Murtaza, B. & Javeed, H. M. R. (2017). Influence of groundwater and wastewater irrigation on lead accumulation in soil and vegetables: implications for health risk assessment and phytoremediation. *Int J Phytoremediation*, 19, 1037–1046 42.

[37] Khalid, S., Shahid, M., Natasha, B. I., Sarwar, T., Shah, A. H. & Niazi, N. K. (2018). A review of environmental contamination and health risk assessment of wastewater use for crop irrigation with a focus on low and high-income countries. *Int J Environ Res Public Health*, 15, 895.

[38] Sohail, M., Eqani, S. A. M. A. S., Podgorski, J., Bhowmik, A. K., Mahmood, A., Ali, N., Sabo-Attwood, T., Bokhari, H. & Shen, H. (2018). Persistent organic pollutant emission via dust deposition throughout Pakistan: spatial patterns, regional cycling and their implication for human health risks. *Sci Total Environ*, *618*, 829–837.

[39] Saeed, M. F., Shaheen, M., Ahmad, I., Zakir, A., Nadeem, M., Chishti, A. A., Shahid, M., Bakhsh, K. & Damalas, C. A. (2017). Pesticide exposure in the local community of Vehari District in Pakistan: an assessment of knowledge and residues in human blood. *Sci Total Environ*, *587*, 137–144.

[40] Khan, M., Mahmood, H. Z. & Damalas, C. A. (2015). Pesticide use and risk perceptions among farmers in the cotton belt of Punjab, Pakistan. *Crop Prot*, *67*, 184–190.

[41] Nayar, K. K., Ananta Krishnan, T. N. & David, B. V. (1989). General and Applied Entomology. Tata McGrow Hill Publishing Co Ltd, New Delhi, India. 17a

[42] Damalas, C. A. & Khan, M. (2016). Farmers' attitudes towards pesticide labels: implications for personal and environmental safety. *Int J PestManag*, *62*, 319–325.

[43] Khooharo, A. A., Memon, R. A. & Mallah, M. U. (2008). An empirical analysis of pesticide marketing in Pakistan. *Pakistan Economic and Social Review*, 57-74.

[44] Saleem, M. & Arshad, M. (2005). *Environmental hazards of pesticides*. Internet www page, at URL: http://www.pakissan.com/english/issues/environmental.hazards.of.pesticides.shtml.

[45] Ullah, S., Hasan, Z., Zorriehzahra, M. J. & Ahmad, S. (2017). Diagnosis of endosulfan induced DNA damage in rohu (Labeo rohita, Hamilton) using comet assay. *Iranian Journal of Fisheries Sciences*, *16*(1), 138-149.

[46] Akhtar, Mobeen., Shahid, Mahboob., Salma, Sultana., Tayyaba, Sultana., Khalid, Abdullah Alghanim. & Zubair, Ahmed. (2014). "Assessment of pesticide residues in the flesh of Catla catla from Ravi River, Pakistan.: *The Scientific World Journal*.

[47] Kalsoom, O. M. M. I. A., Jalali, S. A. M. I. N. A., Shami, S. A., Khan, R., Hamed, M. & Barkati, S. O. H. A. I. L. (2005). Effect of endosulfan on histomorphology of fresh water cyprinid fish, Cyprinion watsoni. *Pakistan Journal of Zoology*, 37(1), 61-67.

[48] Soomro, A. M., Seehar, G. M., Bhanger, M. I. & Channa, N. A. (2008). Pesticides in the blood samples of spray-workers at agriculture environment: The toxicological evaluation. *Pakistan Journal of Analytical & Environmental Chemistry*, 9(1), 6.

[49] Ahmad, Mushtaq. & Iqbal Arif, M. (2008). "Susceptibility of Pakistani populations of cotton aphid *Aphis gossypii* (Homoptera: Aphididae) to endosulfan, organophosphorus and carbamate insecticides." *Crop protection*, 27, no. 3-5, 523-531.

[50] Latif, Y., Sherazi, S. T. H. & Bhanger, M. I. (2011). Monitoring of pesticide residues in commonly used fruits in Hyderabad Region, Pakistan. *American Journal of Analytical Chemistry*, 2(08), 46.

[51] Anwar, T., Ahmad, I. & Tahir, S. (2011). Determination of pesticide residues in fruits of Nawabshah district, Sindh, Pakistan. *Pak J Bot*, 43, 1133–1139.

[52] Food and Agricultural Organization (FAO). FAO/WHO global forum of food safety regulators. Marrakech, Morocco, 28–30 January; 2002. [http://www.fao.org/DOCREP/MEETING/004/AB428E.HTM Agenda Item 4.2 a, GF/ CRD Iran-1].

[53] Randhawa, M. A., Anjum, F. M., Asi, M. R., Butt, M. S., Ahmed, A. & Randhawa, M. S. (2007). *Removal of endosulfan residues from vegetables by household processing*.

[54] Ramesh, A. & Balasubramanian, M. (1999). The impact of household preparations on the residues of pesticides in selected agricultural food commodities available in India, *JAOAC Int*, 82, 725-737.

[55] Yadav, P. R. & Dashad, S. S. (1984). Dissipation of endosulfan residues from unprocessed and processed brinjal (Solanum melongena L.) fruits, *Beitr Trop Land Veterin*, 22, 83-89.

[56] Chavarri, M. J., Herrera, A. & Arino, A. (2004). Pesticide residues in the field- sprayed and processed fruits and vegetables, *J Sci Food Agric*, 84, 1253-1259.

[57] Rehman, A., Jingdong, L., Chandio, A. A., Hussain, I., Wagan, S. A. & Memon, Q. U. A. (2016). Economic perspectives of cotton crop in Pakistan: a time series analysis (1970–2015)(part 1). *J Saudi Soc Agric Sci*.

[58] Latif, Y., Sherazi, S. T. H. & Bhanger, M. I. (2011). Monitoring of pesticide residues in commonly used fruits in Hyderabad Region, Pakistan. *American Journal of Analytical Chemistry*, 2(08), 46.

[59] Muhammad, F., Akhtar, M., Rahman, Z. U., Farooq, H. U., Khaliq, T. & Anwar, M. I. (2010). Multi-residue determination of pesticides in the meat of cattle in Faisalabad-Pakistan. *Egypt. Acad. J. Biol. Sci*, 2, 19-28.

[60] Nafees, M., Jan, M. R. & Khan, H. (2008). Pesticide use in Swat Valley, Pakistan: exploring remedial measures to mitigate environmental and socioeconomic impacts. *Mountain Research and Development*, 28(3), 201-204.

[61] Tariq, M. I., Afzal, S. & Hussain, I. (2004). Pesticides in shallow groundwater of bahawalnagar, Muzafargarh, DG Khan and Rajan Pur districts of Punjab, *Pakistan. Environment International*, 30(4), 471-479.

[62] Ahad, Karam., Tahir, Anwar., Imtiaz, Ahmad., Ashiq, Mohammad., Seema, Tahir., Shagufta, Aziz. & Baloch, U. K. (2000). "Determination of insecticide residues in groundwater of Mardan Division, NWFP, Pakistan: a case study." *WATER SA-PRETORIA*, 26, no. 3, 409-412.

[63] Grondona, S. I., Gonzalez, M., Martínez, D. E., Massone, H. E. & Miglioranza, K. S. (2014). Endosulfan leaching from Typic Argiudolls in soybean tillage areas and groundwater pollution implications. *Sci Total Environ*, 484, 146–153.

[64] Rojas, R., Vanderlinden, E., Morillo, J., Usero, J. & El Bakouri, H. (2014). Characterization of sorption processes for the development of lowcost pesticide decontamination techniques. *Sci Total Environ*, 488-489, 124–135.

[65] De Munari, A., Semiao, A. J. C. & Antizar-Ladislao, B. (2013). Retention of pesticide endosulfan by nanofiltration: influence of organic matter- Environ Sci Pollut Res pesticide complexation and solute-membrane interactions. *Water Res*, *47*, 3484–3496.

[66] Maqbool, Z., Hussain, S., Imran, M., Mahmood, F., Shahzad, T., Ahmed, Z., Azeem, F. & Muzammil, S. (2016). Perspectives of using fungi as bioresource for bioremediation of pesticides in the environment: a critical review. *Environ Sci Pollut Res*, *23*, 16904–16925.

[67] Trinh, H. T., Duong, H. T., Ta, T. T., Van, Cao H., Strobel, B. W. & Le, G. T. (2017). Simultaneous effect of dissolved organic carbon, surfactant, and organic acid on the desorption of pesticides investigated by response surface methodology. *Environ Sci Pollut Res*, *24*, 19338–19346.

[68] Ahmad, Mushtaq. & Iqbal Arif, M. (2008). "Susceptibility of Pakistani populations of cotton aphid Aphis gossypii (Homoptera: Aphididae) to endosulfan, organophosphorus and carbamate insecticides." *Crop protection*, *27*, no. 3-5, 523-531.

[69] Ahmad, Mushtaq. & Shamim, Akhtar. (2013). "Development of insecticide resistance in field populations of Brevicoryne brassicae (Hemiptera: Aphididae) in Pakistan." *Journal of economic entomology*, *106*, no. 2, 954-958.

[70] Saleem, U., Ejaz, S., Ashraf, M., Omer, M. O., Altaf, I., Batool, Z. & Afzal, M. (2014). Mutagenic and cytotoxic potential of Endosulfan and Lambda-cyhalothrin—*In vitro* study describing individual and combined effects of pesticides. *Journal of Environmental Sciences*, *26*(7), 1471-1479.

[71] Akhtar, Shahida., Pirzada, M. H. Siddiqui. & Umar, K. Baloch. (1998). "Effect of neem cake on the persistence of diazinon and endosulfan in paddy soil." *Pesticide science*, *52*, no. 3, 218-222.

[72] Memon, G. Z., Bhanger, M. I. & Akhtar, M. (2009). Peach-nut shells-an effective and low-cost adsorbent for the removal of endosulfan from aqueous solutions. *Pak. J. Anal. Environ. Chem*, *10*(1), 14-18.

[73] Mukhtar, H., Khizer, I., Nawaz, A., Asad-Ur-Rehman. & Ikram-Ul-Haq. (2015). Biodegradation of endosulfan by aspergillus niger isolated from cotton fields of punjab, pakistan. *Pakistan journal of botany*, *47*, 333-336.

[74] Zaffar, H., Sabir, S. R., Pervez, A. & Naqvi, T. A. (2018). Kinetics of Endosulfan Biodegradation by Stenotrophomonas maltophilia EN-1 Isolated From Pesticide-Contaminated Soil. *Soil and Sediment Contamination: An International Journal*, 1-13.

[75] Arshad, M., Hussain, S. & Saleem, M. (2008). Optimization of environmental parameters for biodegradation of alpha and beta endosulfan in soil slurry by Pseudomonas aeruginosa. *Journal of applied microbiology*, *104*(2), 364-370.

[76] Faheem, U., Nazir, T. A. M. S. I. L. A., Saleem, M. A., Yasin, M. & Bakhsh, M. U. H. A. M. M. A. D. (2013). Status of insecticide resistance in Helicoverpa armigera (Hubner) in southern Punjab, Pakistan. *Sarhad Journal of Agriculture*, *29*(4).

[77] Ullah, S., Hasan, Z., Zorriehzahra, M. J. & Ahmad, S. (2017). Diagnosis of endosulfan induced DNA damage in rohu (Labeo rohita, Hamilton) using comet assay. *Iranian Journal of Fisheries Sciences*, *16*(1), 138-149.

In: Endosulfan
Editors: I. C. Yadav and N. L. Devi
ISBN: 978-1-53615-910-3
© 2019 Nova Science Publishers, Inc.

Chapter 7

ENDOSULFAN CONTAMINATION IN SOIL: SOURCES, IMPACT AND BIOREMEDIATION

Somenath Das[1], Anand Kumar Chaudhari[1], Ajay Kumar[2] and Vipin Kumar Singh[1]*

[1]Department of Botany, Centre of Advanced Study, Institute of Science, Banaras Hindu University, Varanasi, India
[2]Agriculture Research Organizations (ARO), Volcani Center, Israel

ABSTRACT

Indiscriminate application of insecticide endosulfan in the agricultural field has polluted the soil, air, and water posing a severe challenging threat to the natural ecosystem. Once introduced into the natural environment they are transformed into different forms depending on the prevailing environmental conditions and microbial community involved. The endosulfan sulfate produced by microbiological activity is considered as one of the stable form responsible for food chain contamination. The endosulfan has a hazardous impact on human and other organisms including fishes, beneficial insects and soil microorganisms due to its persistent, neurotoxic, nephrotoxic, and genotoxic nature. Their presence

---

* Corresponding Author's Email: vipinks85@gmail.com.

in food products beyond the prescribed safety limits has emerged as a challenging issue to human health. The effective detection of endosulfan content using high throughput instrument is very much important for the management of contaminated sites. The presence of endosulfan in groundwater resulting from the surface runoff near the abandoned production site and huge illegal agricultural application in several parts of the world even after the ban indicates the need of strict regulations in the present scenario. Currently, numerous physical and chemical methods have been proposed to eliminate the risk of endosulfan contamination but due to their limitations such as high input of energy, application of toxic chemicals and generation of secondary products of noxious nature has limited the application. On the other hand, biological methods relying on native microorganism being eco-friendly, cost-effective and free from generation of secondary sludge are much preferred. Since the biological processes are largely governed by an environmental factor, optimization of important parameters and selection of suitable microorganism displaying tolerance to a wide range of endosulfan concentration is inevitable. Further, the identification of genes directly involved in degradation and their transfer into suitable microbe may be considered as an effective strategy to control the endosulfan toxicity in the environment.

Keywords: endosulfan, bioaccumulation, genotoxic, biodegradation, insecticide

1. INTRODUCTION

Endosulfan, an organochlorine insecticide is the amalgamation of α- and β- form in the proportion of 7:3 that has been widely used to control the insect pests in different types of crops [1-3]. Chemically, it is the broad spectrum cyclodiene organochlorine insecticide. It is chlorine and sulfur-containing organic compound insoluble in water and accumulated to different trophic level organisms once released in the soil system. Endosulfan may pose a severe threat to the aquatic environment. Endosulfan is one of the recognized persistent organic pollutant (POPs) with gradual accumulation in different trophic levels, low degradation rate and hazardous impact on human health and natural environment [4-5]. Very often they are introduced in larger quantities of vegetables and animals intended for food purposes, thus, posing severe risks [4]. Their long term residence and toxicity to the diverse array of organisms inhabiting in the natural system such as soil and water environment has emerged as a challenging environmental threat, attracting the environmental biologists as well as general public worldwide. In addition to targeting harmful insects, endosulfan may also negatively affect the population of beneficial insects, aquatic organisms and human health. The risks resulting from their large amount presence in the environment is of considerable attention because of endocrine disrupting behavior leading to significant changes in endocrine, and reproductive behavior of humans as well as wildlife.

Prior to 1970s, the POPs had been exploited indiscriminately worldwide [6]. Subsequently, United Nation Environment Program (UNEP) in its Stockholm Convention marked endosulfan as persistent organic pollution [7-8] and banned for field application due to their semi-volatile, persistent nature as well as hazardous impact on environmental and human health. Although many developing and developed nations have decided not to employ the organochlorine pesticides for the intended agricultural or other purposes, its residual toxicity can not be ignored in an affected environmental system [9]. The wide presence of different persistent organochlorine pesticides through distillation effect has been ascribed as one of the major phenomenon for their existence in different regions around the globe [10-11]. In addition, the gradual and uninterrupted loss of these pesticides from agricultural fields has also been considered as an important factor for their release in the natural environment [12]. Endosulfan has recently been detected as the prevalent persistent organic contaminant in urban regions of the central–southern Italian soils [13].

Endosulfan has been demonstrated to pose serious effect on non target organisms including fishes, mammals and other organisms too and the associated impact appears in form of toxicity to reproductive organs, genetic constitution and nervous system [14] because of its persistent nature and long half-life ranging from 60 to 800 days [15].

The endosulfan is endowed with the property of surface sorption onto soil and sediment system because of its hydrophobic nature and results into long term presence in soil and sediment [15-16] as well as in aquatic system [17-18]. Long term availability of endosulfan in aqueous and soil ecosystem culminates into bioaccumulation in crop byproducts [19], aquatic ecosystem components including large aquatic plants [20], phytoplanktons [21], fishes [22], as well as edible herbs and milk derived foods [23]. Considerable differences in the physicochemical nature of endosulfan as compared to other cyclodiene insecticides have a potential impact on its distribution, persistence and microbiological degradation [24]. The α- and β- isomeric forms of endosulfan are susceptible to mineralization via action on sulfite moiety either through oxidation to produce a toxic sulfated form or by hydrolysis giving rise to non-hazardous diol form [25]. The sulfated form of endosulfan has been demonstrated to appear under microbiological actions only while diol form may be synthesized under alkaline conditions [26]. A very little amount of endosulfan and their derived forms, however, is subjected to spontaneous natural removal in an alkaline environment and oxidation by UV irradiation [27].

So far, lots of microbes belonging to bacterial and fungal species have been reported to involve in breakage of endosulfan *in vitro* [25, 28-32]. The key processes participating in biodegradation of endosulfan are oxidation and hydrolysis directing the synthesis of hazardous product endosulfan sulfate and non-hazardous product endosulfan diol, respectively. To combat the adverse influence of endosulfan, several alternative strategies including physical, chemical and biological has been adopted. Many physicochemical treatments like flocculation, photooxidation, reduction etc. are employed to degrade pesticides but these treatments are not so effective [33]. Thus, degradation of complex toxic endosulfan into simple non toxic form by biological organisms (especially microorganisms) is highly recommended [34]. Several studies have addressed the importance of microorganisms such as bacteria, fungi, actinomycetes and others for the degradation of harmful endosulfan. Microorganisms break this harmful compound and use them as a source of a substrate to gain energy. The present chapter has been designed with the aim to discuss the different aspects of endosulfan contamination in soil, impact on groundwater pollution, toxicity, the effect on vital plant physiological processes along with remediation using physicochemical and biological techniques.

2. ENDOSULFAN CONTAMINATION IN SOIL

Application of endosulfan against the insect pests has long been practiced for the yield management of several plants including vegetables, fruits, and vegetables [35]. Despite the ban on endosulfan, the widespread application of broad-spectrum endosulfan in many developing nations to control the losses of crop productivity has aggravated the contamination of soil [36-38]. Detection of different pesticides along with endosulfan using ultra-sonication technique coupled with GC-MS analysis revealed the presence of fairly high content of endosulfan and endosulfan sulfate in soil from a horticulture area in Portugal [39]. The β-endosulfan was the most prevalent isomer present in contaminated soil. They have reported the decrease in the content of endosulfan with increasing soil depth. Interestingly, the content of sulfated endosulfan was higher than the cumulative content of α- and β- isomeric forms of endosulfan, suggesting the biological degradation of endosulfan due to bacterial activities resulting into the synthesis of endosulfan sulfate, one of the noxious and biodegradation resistant forms. Soil and water contamination from various agriculture areas of Haryana, India with different pesticides including endosulfan has been described [40]. They have described the soil endosulfan level ranging from 0.002–0.039 $\mu g\ g^{-1}$ with the dominance of β- form. Presence of different levels of endosulfan with the prevalence of endosulfan sulfate from seven cities of India including New Delhi, Agra, Kolkata, Mumbai, Goa, Chennai, and Bangalore has been reported [41-42]. High dominance of endosulfan was attributed to the high stability to degradation. Endosulfan content was recorded to range from 0.01 and 237 ng/g dw with the highest content in Goa. The ratio of α-/β-endosulfan was observed to reach up to 2.33 suggesting the continuous exploitation of persistent organochlorine pesticides. Endosulfan contamination in the soil of Kasargode, Kerala is also described [43]. Contamination of endosulfan was observed because of recurrent spraying of the pesticides on cashew plantations approximately for twenty years. Soil contamination with endosulfan had the total content ranging from 0.001–0.010 $\mu g\ g^{-1}$. Presence of endosulfan in urban soil of Nepal serving as a secondary point of emission is recently claimed by Pokhrel et al. [44]. The total endosulfan content (α- and β- isomeric form) in soil was recorded in between 0.01 to 16.4 ng/g dw. The higher flux for endosulfan was noticed for Kathmandu region as compared to Pokhara region of Nepal. Presence of endosulfan in soil at different depth nearby the endosulfan production site in Jiangsu, China has been evaluated. The content of soil endosulfan varied from 0.01 to 114 mg/kg d.w. at the production site while it was observed to range from 1.37–415 ng/g d.w. in the soil of nearby areas indicating the exposure and hence health risks to the exposed population. There were variations in the content of endosulfan at the surface of the soil system whereas, after the depth of 120 cm, the decline in total quantity was found.

3. SOURCES OF ENDOSULFAN AND ITS DETECTION TECHNIQUES

The major sources of endosulfan include their agricultural application to control the insect pest infesting cotton, vegetables, fruits, tea, coffee and cereals [38, 45-47]. Endosulfan contamination also results from their application for controlling the mites [48], ultimately reaching to water and soil system. In addition to this, the industrial areas previously synthesizing endosulfan are an important source of endosulfan in soil [49]. Surface run-off, accidental spills and leaching of endosulfan from these abandoned production units has lead to contamination of nearby sites of agricultural importance [50-51]. Furthermore, agricultural soils are also susceptible to endosulfan contamination by atmospheric deposition of globally distributed semi-volatile organochlorine pesticides i.e., endosulfan.

So far research on endosulfan has used various analytical techniques for its estimation in the contaminated system. The currently used estimation techniques are relied on high-performance liquid chromatography (HPLC), gas chromatography-mass spectrometry (GC-MS), and colorimetric methods. However, HPLC and GC-MS method is considered relatively more appropriate for quantification of endosulfan in contrast to colorimetric methods. Prior to estimation, the contaminated soil is processed to extract the endosulfan followed by its quantification.

Raju and Gupta [52] have described the colorimetric methods for the determination of endosulfan. The method is relied upon the liberation of sulfur dioxide gas followed by its absorption in malonyl di hydrazide. The resulting solution was finally mixed with an acidic solution of p-aminoazobenzene and formaldehyde leading to the formation of the pink colored product. The colored product had its absorption maxima at 505 nm wavelength. The proposed method obeyed the Lambert-Beer's law in the range of 1-6 µg/mL. The method was quite simple and free from interferences generally caused by naturally occurring ions and contaminants. They reported the method's suitability to detect as low as 0.25 µg mL^{-1} endosulfan in soil samples.

Pasha and Narayana [53] have demonstrated the application of chromogen thionin and methylene blue for the colorimetric determination of endosulfan. This method was also based on the release of sulfur dioxide gas which was passed subsequently in the solution of potassium iodate causing the emergence of iodine. The discharged iodine was found to bleach the color of thionin and methylene blue which can be measured through spectrophotometer at 600 nm and 665 nm, respectively. The reduction in optical density was linearly correlated with the quantity of endosulfan. The linearity was observed in between 0.4–7.0 and 0.2–9.0 µg mL^{-1} endosulfan for thionin and methylene blue, respectively. The proposed technique was tested for soil system and not affected by the presence of different ions present in natural water or soil samples.

Another method for spectrophotometric measurement of endosulfan was demonstrated by Venugopal and Sumanlatha [54]. The introduction of alcoholic potassium hydroxide solution in sample resulted in the discharge of sulfur dioxide. The generated gas after refluxing with H_2O_2 and diphenylamine gives the colored product with absorption maxima at 605 nm wavelength. The linearity of the calibration curve was observed in the concentration range of 0.9 to 5.0 ppm. The suitability of developed methodology has been claimed for soil and free from interferences resulting from common contaminants.

4. IMPACT OF ENDOSULFAN CONTAMINATION IN SOIL

4.1. Effect of Soil Endosulfan Contamination on Surface and Groundwater

Globally, there has been rising evidence of indiscriminate application of agrochemicals [55] with diverse chemistry causing environmental contamination [56]. The huge and extensive application has deteriorated the quality of water, soil as well as the atmosphere [57]. Since, pesticides after application may get access to different aquatic ecosystem including groundwater and surface water through surface runoff [40] during rainy season, desorption, leakage from contaminated containers and instrument's cleaning processes [58], their presence may have serious consequences on freshwater reserves and thus on human health. The pesticide transfer from soil into aqueous environment is affected by different environmental conditions. Most of the applied pesticides, find their ways into the nearby environment causing an ecological imbalance [59]. However, a systematic study concentrating on the estimation of various pesticides in groundwater at the global level is still in the incipient phase [60]. The geographical regions where groundwater serves as the major sources of drinking water, pesticide-contaminated aquifers may pose severe risks to inhabiting populations [61]. Pollution of groundwater by hazardous synthetic pesticides has multiple negative impacts on natural ecosystems. Groundwater system contaminated by pesticides generally remains polluted for very long duration because of resistance against degradation and may have serious risks of bioaccumulation too [62].

Endosulfan has recently been detected as one of the important contaminants of groundwater in northern Lebanon [63]. The concentrations of sulfated form at different locations were higher in contrast to the non-sulfated form of endosulfan. The maximum concentration of endosulfan was recorded as 0.08 µg/L while that for endosulfan sulfate was observed as 1.65 µg/L. Different organochlorine pesticides including endosulfan II were found as the contaminant in the surface as well as groundwater nearby the Pampanga river of Philippines [64]. The concentration in surface water ranged from 0.04 μg/L to 0.064 μg/L whereas in groundwater the contaminant level was in between 0.047 μg/L to 0.074 μg/L.

Their presence in the aquatic environment beyond the WHO acceptable limit is an emerging environmental challenge worldwide and suggests indiscriminate and illegal application in agro-ecosystem in spite of their ban in most of the countries across the globe. Hence, regular interval assessment is necessary in order to get the status of pesticides in the aquatic ecosystem for proper management of polluted sites. Recently, Pan et al. [60] have demonstrated the mean level of total endosulfan in groundwater samples of Yangtze River Basin reaching up to 244.44 ng/L and found its widespread contamination in the investigated region. The presence of endosulfan in groundwater was attributed to the shallow depth as well as non-degradable nature of pollutant. Bai et al. [65] have described the concentration of endosulfan I in surface and groundwater of Shaying River located in China, as 8.5 and 5.5 ng/L, respectively. The presence was highly associated with the detection of heptachlor, heptachlor oxide and α-HCH indicating simultaneous application. The endosulfan was encountered as the major surface water contaminant of the investigated area.

4.2. Effect of Endosulfan on Human Health

Toxicity of endosulfan to reproductive organs, kidney and immune system in human and mammals is well established [66-71]. Zhang et al. [72] have elucidated the toxicological impacts of endosulfan in the induction of oxidative stress and concluded the risk of development of cardiovascular disorders. Endothelial cells after treatment with varying concentrations of endosulfan were noticed to activate the phenomenon of autophagy. Treatment with endosulfan resulted into diminished ATP pool and membrane potential in a dose-dependent way. Elevations in inflammation and expression after endosulfan treatment were observed for tumor necrosis factor (TNF), chemotactic proteins and biomolecules facilitating in cellular communications. Recently, endosulfan induced infertility in males caused by the qualitative and quantitative reduction in sperm has been reported by

Sebastin and Raghvan [73]. Endosulfan treatment had a potentially negative impact on the process of spermatogenesis leading to significant decrease in spermatids. Role of endosulfan in restriction of normal cell cycling processes, provoking oxidative stress, enhanced secretion of lactate dehydrogenase, inhibition of cell division, degeneration of cytoskeleton components and induction of apoptosis in endothelial cells has been proposed recently [74]. Endosulfan was demonstrated to modulate the Notch signaling pathway by altering the synthesis of protein participating in the pathway. Lipophilic nature of endosulfan makes its accessibility and hence accumulation to bone marrow. Higher concentrations of endosulfan in bone marrow of children dwelling near the contaminated sites and prevalence of blood cell cancer as compared to those inhabiting in a pollutant-free area is evidenced [75]. Association of endosulfan with increased probability of leukemia is also presented [76]. Treatment with different rising concentrations of endosulfan to K-562 cell lines resulted into elevated loss of cellular viability through loss of DNA stability with enhanced formation of micronuclei at higher concentration. Treatment was found to up-regulate the genes governing the synthesis of p53 and GADD45A while down-regulation was observed for PCNA and XRCC2.

4.3 Effect of Endosulfan in Plants

Huge application of few pesticides including endosulfan has improved the crop productivity globally, however; the heavy accumulation in soil, air, water, vegetables, grains, fruits, feed, and other commodities cannot be avoided. Laboratory, studies have shown that endosulfan imposes a negative effect on vital plant physiological processes. Endosulfan and its sulfated form induced decrease in root/shoot growth and fresh weight, chlorophyll content, root/shoot tolerance index and enhancement in malondialdehyde (MDA) content, an indication of elevated oxidative stress in a dose-dependent manner is described [3, 77-78].

Generally, the decrease in root and shoot growth is linked with the reduced productivity of crops [79]. Most of the synthetic pesticides alter plant growth by restricting cell division, expansion, cellular differentiation, seed germination and development of seedlings [80]. The negative impact of endosulfan in terms of reduced growth, photosynthetic pigments, and enhanced oxygen evolution and respiration in *Azolla* have been elaborated by Waseem and Preeti [81]. Significant induction in the level of the antioxidant enzymatic system, as well as antioxidants after application of endosulfan at different time intervals to minimize the hazardous effect of pesticide on vital cellular processes, is evidenced in the investigations carried out on *Cajanus cajan* [82]. Enhancement in catalase, glutathione-S-transferase, glutathione reductase and H_2O_2 content in macrophyte *Myriophyllum quitense* exposed to varying concentrations of endosulfan implying the induction of oxidative stress is also presented [50]. Kumar [83] have demonstrated the detrimental effects of endosulfan application on cyanobacterial species including *Aulosira fertilissima*, *Anabaena variabilis* and *Nostoc muscorum* commonly occurring in the paddy field. The decrease in important pigments, sugar, cellular multiplication together with enhancement in total protein, osmolytes, malondialdehyde and important defense enzymatic system such as catalase, ascorbate peroxidase, and superoxide dismutase was apparently documented. Such alteration may have negative consequences on soil fertility and therefore crop productivity. Similarly, the impact of endosulfan on cyanobacterium *Plectonema boryanum* and species of *Nostoc* is demonstrated by [84-85]. Recently, Rani et al. [86] have presented the effect of increment of endosulfan content in *Helianthus annuus* on malondialdehyde content synthesized in leaves and roots, implicating the induction of cellular oxidative stress. The decrease in root/shoot length, dry weight, photosynthetic pigments and protein content with increase in endosulfan was observed in a dose-dependent manner. Harmful impacts of endosulfan application on plants in terms of chromosomal abnormality, micronuclei formation, and reduced seedling growth have also been reported earlier [87-88].

Such detrimental impacts on important biochemical processes have been suggested to limit the productivity of several important crops. Enhanced phenolic content of leaves has been noticed after rice plantations treatment with endosulfan in order to control the losses caused by insect infestations [89]. Although, there is a number of publications on pesticide toxicity and their impact on crop physiological processes [90] still farmers are indiscriminately using the banned pesticides. Once introduced into the environment, the pesticides have hazardous impacts not only on the soil biological and physicochemical properties but also on the nutritional status of different crops.

4.4. Effect of Endosulfan on the Microorganism

Microorganism represents an important part of the soil food chain and food web; they have been described to maintain vital pools of organic matter in soil ecosystem functioning. Presence of different soil dehydrogenase enzymes serve as an indicator of soil microbial diversity and their oxido-reductive process. Nowadays increase in agricultural productivity in developing countries is fully dependent on the use of different synthetic insecticides and pesticides. Underdeveloped countries are still using different synthetic pesticides which are harmful to the human health and surrounding atmosphere. Organochlorines are one of these synthetic pesticides, currently regarded as environmental pollutant throughout the world. Endosulfan is a broad spectrum organochlorine cyclodiene pesticides widely used for controlling different insect pests of coffee, cotton, paddy, oilseeds, and vegetables [91]. However, the wide use of endosulfan causes a major risk for soil as well as water pollution. Continuous use of endosulfan affects the soil microorganisms and their normal biological activities. Endosulfan is reported to inhibit a large number of methanogenic bacteria in soil. Iqbal et al. [92] showed rapid decrease in soil fungal population by continuous application of endosulfan.

Toxic effects of endosulfan on soil microorganisms cause reduction of dehydrogenase, β-glucosidase, and alkaline phosphate activity [93]. Endosulfan treatment also caused a decrease in phosphorus availability in soil due to inhibition of phosphate solubilizing activity of numerous bacteria. It also acts as a neurotoxic agent that directly bind with γ aminobutyric acid (GABA) linked chloride ion channel in the central nervous system [94]. Among the diastereomeric mixture of α- and β- endosulfan, β-isomer is rapidly hydrolyzed in water and causes acute toxicity to fishes and other invertebrates in the aquatic ecosystem. The lower concentration of endosulfan has been reported to decrease absorption of different minerals *viz.* Ca, Mg, Co, and Cu from soil [95]. It is also responsible for changes in body movement, nervous system and leads to paralysis or even death in *Barylelphusa cuniculris* [96]. Chaudhari et al. [97] and ASTDR [98] have reported mutagenic effect of endosulfan on bacterial and yeast cell system. Further, increased application of endosulfan (200 g/ha) in the field has been reported to cause immature death of frogs, snakes, and lizards [99].

5. REMEDIAL MEASURES FOR ENDOSULFAN CONTAMINATION

5.1. Physico-chemical Methods of Endosulfan Degradation

Endosulfan contamination from various point and non-point sources such as agricultural runoff, municipal wastes, and pesticide industries is a major problem to an ever-increasing population of the current generation. It may easily contaminate groundwater, surface water, large aquatic and soil ecosystem. Lipid solubility of endosulfan leads to rapid accumulation of endosulfan in aquatic organisms and further biomagnifications through food chain. Therefore, cumulative interest has focused on the removal of this pesticide from soil, surface water, and groundwater.

There are different physicochemical methods *viz.* absorption, incineration and oxidation for successful elimination of endosulfan contamination. Yonli et al. [100] reported the successful removal of α-endosulfan from water by HY and steamed HBEA zeolites. Currently, advanced oxidation process (AOP) employing reactive hydroxyl radical has emerged as an efficient alternative for removal of numbers of hazardous pollutants from soil and water [101]. Reverse osmosis, sand filtration, chemical oxidation, and carbon absorption are important physical techniques for partial removal of organochlorine pesticides from contaminated water. Liquid-liquid extraction (LLE) and solid phase extraction (SPE) is another major physicochemical process for extraction of endosulfan from environmental samples. The efficiency of extraction is dependent on the chemical properties of endosulfan such as hydrophobicity, vapor pressure, molecular weight, and solubility. Dechlorination of toxic endosulfan to less toxic intermediates by nano-zero valent iron (ZVI) and zero-valent magnesium (Mg^0) was observed by Singh et al. [102]. Begum et al. [103] observed the dechlorination of aqueous phase endosulfan through $Mg^0/ZnCl_2$ metallic hydrocarbon.

The powdered form of Zn^0 has demonstrated the potential capability to degrade soil endosulfan through reduction of chloride groups under acidic conditions [104]. Advanced chemical technology for endosulfan degradation includes ozonation with specific electrophilic, nucleophilic and dipolar reaction as well as hydroxylation reaction. Optimization of various parameters including ozone concentration, pH stabilization, temperature maintenance, and kinetic studies was done for optimal endosulfan degradation [103]. Gupta et al. [105] have reported adsorption of the toxic endosulfan on activated carbon as one of the most widespread technology for removal of endosulfan.

Endosulfan and its by-products could also be degraded by sunlight as well as xenon arc lamp [106]. Shah et al. [107] indicated oxidative system activated with iron compounds leading to quick degradation of aqueous endosulfan by photocatalysis. Xiong et al. [108] reported that positively charged TiO_2 surface is actively involved in surfactant elution and electrostatic attraction for efficient photocatalytic degradation of endosulfan. Some of the important studies pertaining to physicochemical degradation of endosulfan are presented in Table 1.

Table 1. Physicochemical methods for the removal of endosulfan

Techniques used	Techniques used for degradation assessment	Remarks	References
Advanced oxidation process (Non-thermal plasma integrated with cerium oxide catalysts)	HPLC with photo diode array (PDA) detector, TOC analysis; gas analyzers	Degradation followed the first order kinetics; degradation was proportional to applied electricity	[133]
Adsorption (use of carbon produced from generators)	GC	Fast adsorption within 90 minutes; adsorption followed pseudo-second order reaction	[105]
Electrodialysis	Scintillation counter for radio-labeled endosulfan and GC equipped with electron capture detector (ECD)	Efficient binding of endosulfan to ion exchange membrane; sorption was high at neutral pH	[134]
Surfactant addition to contaminated sites	GC equipped with ECD	Biological surfactant was more effective as compared to synthetic ones; surfactant addition into the soil above critical miceller concentration enhanced the removal of endosulfan	[135]
Advanced oxidation based on UV irradiation and	GC and GC-MS	Irradiation was effective in presence of per sulfate ion; organic matter and common anions had inhibitory effect on removal	[136]
Photo-catalytic degradation using silver doped TiO_2 nanocrystal	-	Highly efficient degradation by Silver doped TiO_2 as compared to free form of TiO_2 nanocrystal	[137]

Table 1. (Continued)

Techniques used	Techniques used for degradation assessment	Remarks	References
H_2O_2, ultra-sonication and combination of H_2O_2 and ultra-sonication	GC-MS	The combined efficiency was higher as compared to H_2O_2 alone after 60 minutes; most of the pesticides were completely degraded; total organic carbon (TOC) removal of selected process was assessed at different pH, H_2O_2 dose, pesticide concentration, and time	[138]
Photo-catalytic degradation	HPLC equipped with PDA detector	Photo-catalytic degradation was affected by pesticide content, pH and catalyst content; nearly 100 per cent degradation was achieved	[139]
Atmospheric air cold plasma	GC-MS	Degradation followed first order kinetics; process was efficient in degrading 57% of endosulfan; degradation product was less toxic as compared to parent molecule	[140]
Zero valent iron (ZVI); Mg^0	GC and GC-MS	Nano sized ZVI were more effective than micro sized ZVI; formed by-products after degradation were somewhat de-chlorinated; chlorine removal was sequential	[102,141]
Fenton's reagent	GC-MS	Degradation followed first order kinetics; step by step degradation followed by formation of methyl cyclohexane and 1-hexene was observed	[142]

5.2. Biological Methods of Endosulfan Degradation

Most of the synthetic pesticides used for agricultural purposes act as the xenobiotic substances. Currently, the environmental fate of toxic xenobiotic, endosulfan is a major point of focus. It is reported that among different stereoisomers of endosulfan, β-isomer is more persistent and causes considerable toxicity to different metabolic processes. Another notable character of endosulfan is its biotransformation into endosulfan sulfate, having toxicity similar to parent compound and further degradation is very limited. In anaerobic conditions, *viz.* sediments, and sludges, endosulfan is metabolized into endosulfan diol, a comparatively less harmful product as compared to endosulfan itself [109]. There are different types of degradation system (biotic and abiotic) of endosulfan in soil and water depending upon the solubility, diffusion and sorption pattern. Endosulfan degradation through biological routes has received popular attention because biodegradation process generally includes heterotrophic microorganisms which can degrade the endosulfan easily to fulfill their carbon, nutrient and energy requirements. Several reports exhibited degradation of toxic endosulfan by commonly occurring environmental microorganisms. Bhalerao et al. [110] observed the ability of fungal isolate *Aspergillus niger* in the degradation of toxic endosulfan. Martens [26] reported endodiol and endosulfan as the major metabolite after degradation of endosulfan by bacteria and fungi, respectively. A hypothetical pathway of endosulfan metabolism to endosulfan, endodiol, endohydroxy ether and endolactone was described by Miles and Moy [111]. Hydrolysis of endosulfan into endosulfan diol causes less toxicity to fish and other aquatic organisms. Further degradation process leads to formation of non toxic intermediates. Strain CS5 of *Achromobacter xylosoxidans* is suggested to utilize endosulfan as a sole source of carbon, sulfur, and energy [112]. Mixed bacterial culture of *Stenotrophomonas maltophilia* and *Rhodococcus erythropolis* can easily degrade endosulfan contaminated agricultural soil due to a synergistic effect of a bacterial mixture exhibiting wide range of substrate specificity [113]. Application of mixed bacterial culture having the ability to remove the sulfur group from endosulfan causing decreased vertebrate toxicity due to concurrent detoxification is well described [24]. *Fusarium ventricosum* has been described to actively participiate in degradation of endosulfan into less toxic endosulfan diol [14, 32]. There is a group of fungi such as *Mucor thermohyalospora*, *Aspergillus niger*, *Trichoderma harzianum* and *Phanerochaete chrysosporium* that may utilize endosulfan as their carbon and energy substrate. Kullman and Matsumura [29] reported the active participation of hydrolytic and oxidative enzymes in degradation of endosulfan as a substrate by white rot fungus, *Phanerochaete chrysosporium*.

A number of bacterial, fungal and soil-borne populations of microbes are known to be involved in the degradation of endosulfan and utilization of endosulfan as a source of carbon and sulfur or both [14, 32]. The rate of degradation depends on favorable soil temperature, pH, moisture, and so forth [114]. Kumar and Philip [115] reported the capability of *Staphylococcus* and *Bacillus circulans* towards endosulfan degradation. Kwon et al. [27] described the biodegradation of endosulfan by *Klebsiella oxytoca*. Bacterial degradation of both α- and β- endosulfan by *Pseudomonas spinosa*, *P. aeruginosa*, and *Burkholderia cepacia* was also reported by Hussain et al. [45]. *Stenotrophomonas maltophilia* a bacterium found in the gut of cockroach was represented to metabolize both α- and β- endosulfan after exposure to high level of endosulfan as challenge pesticide [116]. Besides bacteria, some fungi like *Aspergillus niger* plays an important role in mineralization of endosulfan to CO_2 which is utilized as whole CO_2 source to the fungus [110].

Katayama and Matsumura [117] reported the metabolic degradation of endosulfan by *Trichoderma harzianum* into non-toxic form under nutritional supply throughout the growth. Two species of *Aspergillus* (*Aspergillus terrius* and *A. terricola*) and one species of *Chaetosartorya* (*Chaetosartorya stromatoides*) has been investigated for their excellent capability to degrade both α- and β- form of endosulfan [46]. Along with aforementioned group of microorganisms, some green algae like the species of *Chlamydomonas*, *Chlorella*, *Chlorococcum* and *Scenedesmus* [16, 118], blue-green algae, mainly the species of *Anabaena* and *Nostoc* [119] and actinomycetes such as *Arthrobacter*, *Clavibacter*, *Nocardia*, and *Brevibacterium* [120, 121] have also been reported to play a crucial function in biodegradation of toxic endosulfan. Biological degradation of endosulfan by different groups of microorganisms is presented in Table 2.

Table 2. Biodegradation of endosulfan by some important microorganisms

Microorganism types	Microorganism	Detection techniques used for degradation	References
Bacteria	*Klebsiella oxytoca*	GC	[27]
	Klebsiella pneumonae	GC-MS	[143]
	Acinetobacter sp.	GC-MS	[143]
	Flavobacterium sp.	GC-MS	[143]
	Micrococcus sp.	GC-MS	[144]
	Pandoraea sp.	HPLC	[32]
	Ochrobacterum sp.	GC-MS	[145]
	Arthrobacter sp.	GC-MS	[145]
	Burkholderia sp.	GC-MS	[145]
	Bacillus spp.	GC-MS	[30]
	Mycobacterium strain ESD	GC-MS	[25]
	Bordetella sp. B9	GLC	[146]
	Achromobacter xylosoxidans	HPLC	[112]
	Pusillimonas sp. JW2 and *Bordetella petrii* NS	GC-MS	[147]
	Pseudomonas sp. strain TAH	GC-MS	[148]
Fungi	*Fusarium ventricosum*	HPLC	[32]
	Trichoderma harzianum	GLC	[117]
	Mucor thermohyalospora MTCC1384	GC-MS	[149]
	Aspergillus terreus	GLC	[150]
	Aspergillus niger	GLC	[28]
	Cladosporium oxysporum	GLC	[150]
	Phanerochaete chrysosporium	GC-MS	[29]
	Aspergillus and *Trichoderma* spp	GC	[151]
	Botryosphaeria laricina JAS6, *Aspergillus tamarii* JAS9 and *Lasiodiplodia* sp. JAS12	HPLC	[152]
Microalgae and cyanobacteria	*Scenedesmus* sp.	GC, GC-MS	[16]
	Anabaena sp.	GC	[132]
	Chlorococcum sp	GC, GC-MS	[16]
	Chlorella vulgaris	GC	[153]
	Chlorococcum sp. or *Scenedesmus* sp.	GC, GC-MS	[16]
	Westiellopsis prolifica, Nostoc hatei, Anabaena sphaerica	GC	[154]

5.2.1. Mechanism of Biological Degradation

Numbers of studies have reported the mechanism of endosulfan degradation by different microbes. These microbes possess the capability to associate physicochemically with substrate causing significant alterations in structure or absolute breakage of the selected compound. Amongst large pool of microorganisms, bacterial and fungal species are the most widely investigated endosulfan degrader [122], that degrade the endosulfan into less toxic forms (endosulfan sulfate, endosulfan lactones, dieldrin, and octane) by secreting hydrolytic enzymes [123]. Many bacteria like *Stenotrophomonas maltophila, Klebsiella pneumoniae, Pseudomonas* sp. and fungi mainly the species of *Aspergillus, Fusarium,* and *Trichoderma* utilize endosulfan as a exclusive resource of sulfur and carbon and convert it into less toxic endosulfan sulphate, endosulfan diol, endosulfan lactone, endosulfan hydroxy carboxylate, dieldrin and octane [116]. Endosulfan diol is the major metabolite of bacterial actions; however, fungi may utilize endosulfan lactone as the major metabolite source [124]. The biotransformation of endosulfan and other pesticides into less toxic form mainly involves three basic steps (1) metabolism through oxidation, reduction, and hydrolysis, (2) conjugation of resulting residues with sugar and amino acids to enhance solubility and lessen toxicity and, (3) further conversion of second step residues into least toxic forms. The probable mechanism underlying the degradation of endosulfan by major groups of bacteria and fungi is diagrammatically presented in Figure 1.

Various genomic and proteomic approaches have also been used to determine the expression of a particular protein responsible for the degradation of pesticides. Some workers have reported that the expression of gene varies with external condition [125]. For instance, few bacteria under normal condition synthesize extracellular enzymes to degrade these pesticides, however, under stress condition the expression of those particular enzymes is quite different. Some important bacterial genes like opd from *Pseudomonas diminuta,* adpB from actinomycete *Nocardia,* ophB from *Burkholderia,* Mpd from azeotropic *Rhizobium radiobacter* and moulds gene such as A-opd from black mould *Aspergillus niger* and P-opd from blue mould *Penicillium* sp. have been widely investigated for their expression toward pesticide degradation [126-128].

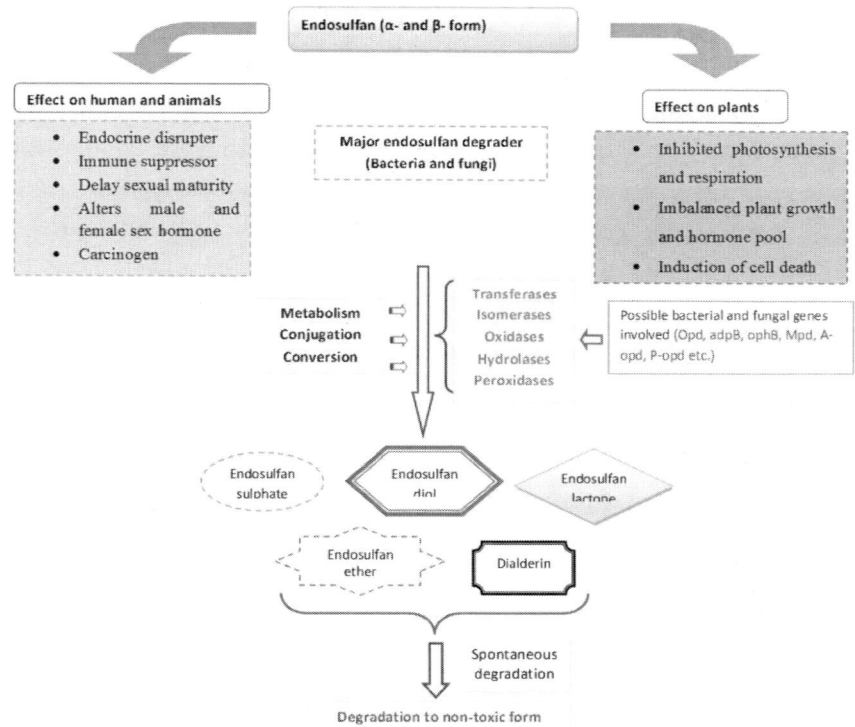

Figure 1. Diagrammatic representation of possible mechanism of microorganism mediated endosulfan degradation.

5.2.1.1. Factors Affecting Biological Degradation of Endosulfan

The rate of endosulfan degradation by microorganisms exhibits a high degree of variability, however, some important factors like favorable moisture content, pH, presence or absence of carbon, biomass, oxygen, amount of microbial population and concentration of pesticides during degradation process may regulate the overall degradation phenomenon [129].

Moisture content is one of the important abiotic parameters regulating the persistence and degradation of endosulfan in the soil. Till date, most of the studies conducted on endosulfan degradation by microorganisms have described the requirement of at least 30% moisture content [115]. The rate of degradation was enhanced in soil under waterlogged condition as compared to water free soil as reported by Awasthi et al. [130]. Some studies also suggested the rate of degradation was optimum at 60% moisture content. The favorable moisture and aeration in the soil may regulate the activity of soil microbes that can cause quick and faster degradation. The degradation rate also depended on the season, as in spring and summer, the degradation was faster as compared to autumn and winter [131]. Like moisture content, pH of the soil also contributes significantly to the degradation of endosulfan, most probably, under alkaline pH, the rate of degradation is optimum. Most of the fungi facilitate rapid degradation of endosulfan at acidic pH [132]. Moreover, carbon also plays an important role in the endosulfan degradation process. Awasthi and [130] and Guerin [109] described that the degradation of endosulfan was enhanced with an external supply of carbon but to a certain limit. These are some important aspects, which must be taken into consideration for the successful and rapid degradation of endosulfan by microbiological processes.

SUMMARY AND CONCLUSION

Endosulfan is one of the toxic insecticides extensively used for management of crop production globally. The applied endosulfan has found its destination into soil, water, and air due to its persistent nature. The hydrophobic nature is capable to interrupt the multitude of vital cellular function and create the oxidative stress beyond the safety limits. The appearance of various human health disorders associated with endosulfan application has raised the serious concern about human health. The high level of endosulfan in soils, air, vegetables, fishes, grains and groundwater has necessitated the development of effective green technology not only for detection of even minute quantity in an environment but also for the elimination of highly noxious levels.

REFERENCES

[1] Weber, Jan, Crispin J. Halsall, Derek Muir, Camilla Teixeira, Jeff Small, Keith Solomon, Mark Hermanson, Hayley Hung, and Terry Bidleman. 2010. "Endosulfan, a global pesticide: a review of its fate in the environment and occurrence in the Arctic." *Science of the Total Environment* 408:2966-2984.

[2] Roberts, Darren M., Ayanthi Karunarathna, Nick A. Buckley, Gamini Manuweera, M. H. Sheriff, and Michael Eddleston. 2003. "Influence of pesticide regulation on acute poisoning deaths in Sri Lanka." *Bulletin of the World Health Organization* 81:789-798.

[3] Singh, V., A. Lehri, and N. Singh. 2018a. "Assessment and comparison of phytoremediation potential of selected plant species against endosulfan." *International Journal of Environmental Science and Technology*: 1-18.

[4] Nakata, H., M. Kawazoe, K. Arizono, S. Abe, T. Kitano, H. Shimada, W. Li, and X. Ding. 2002. "Organochlorine pesticides and polychlorinated biphenyl residues in foodstuffs and human tissues from China: status of contamination, historical trend, and human dietary exposure." *Archives of Environmental Contamination and Toxicology* 43:0473-0480.

[5] Jones, Kevin C., and P. De Voogt. 1999. "Persistent organic pollutants (POPs): state of the science." *Environmental Pollution* 100:209-221.

[6] Wong, M. H., A. O. W. Leung, J. K. Y. Chan, and M. P. K. Choi. 2005. "A review on the usage of POP pesticides in China, with emphasis on DDT loadings in human milk." *Chemosphere* 60:740-752.

[7] World Wide Found, Stockholm Convention. April 2005. "New POPs": screening additional POPs candidates, 38 pp.

[8] Xu, Weiguang, Xian Wang, and Zongwei Cai. 2013. "Analytical chemistry of the persistent organic pollutants identified in the Stockholm Convention: A review." *Analytica Chimica Acta* 790:1-13.

[9] Weinberg, J. 1998. "Overview of POPs and need for a POPs treaty." In *Public forum on persistent organic pollutants-the international POPs elimination network*.

[10] Szeto, Sunny Y., and Patricia M. Price. 1991. "Persistence of pesticide residues in mineral and organic soils in the Fraser Valley of British Columbia." *Journal of Agricultural and Food Chemistry* 39:1679-1684.

[11] Kim, Jong-Hun, and Alistair Smith. 2001. "Distribution of organochlorine pesticides in soils from South Korea." *Chemosphere* 43:137-140.

[12] Meijer, S. N., M. Shoeib, L. M. M. Jantunen, Kevin C. Jones, and T. Harner. 2003. "Air− soil exchange of organochlorine pesticides in agricultural soils. 1. Field measurements using a novel in situ sampling device." *Environmental Science and Technology* 37:1292-1299.

[13] Thiombane, Matar, Attila Petrik, Marcello Di Bonito, Stefano Albanese, Daniela Zuzolo, Domenico Cicchella, Annamaria Lima, Chengkai Qu, Shihua Qi, and Benedetto De Vivo. 2018. "Status, sources and contamination levels of organochlorine pesticide residues in urban and agricultural areas: a preliminary review in central–southern Italian soils." *Environmental Science and Pollution Research* 25:26361-26382.

[14] Siddique, Tariq, Benedict C. Okeke, Muhammad Arshad, and William T. Frankenberger. 2003a. "Enrichment and isolation of endosulfan-degrading microorganisms." *Journal of Environmental Quality* 32: 47-54.

[15] Rao, D. Mohana Ranga, and A. Satyanarayana Murty. 1980. "Persistence of endosulfan in soils." *Journal of Agricultural and Food Chemistry* 28, no. 6: 1099-1101.

[16] Sethunathan, Nabrattil, M. Megharaj, Z. L. Chen, Brian D. Williams, Gareth Lewis, and Ravendra Naidu. 2004. "Algal degradation of a known endocrine disrupting insecticide, α-endosulfan, and its metabolite, endosulfan sulfate, in liquid medium and soil." *Journal of Agricultural and Food Chemistry* 52:3030-3035.

[17] Witter, J. V., Dwight E. Robinson, Ajai Mansingh, and K. M. Dalip. 1999. "Insecticide contamination of Jamaican environment." V. Island-wide rapid survey of residues in surface and ground water." *Environmental monitoring and assessment* 56, no. 3: 257-267.

[18] Chusaksri, S., S. Sutthivaiyakit, and P. Sutthivaiyakit. 2006. "Confirmatory determination of organochlorine pesticides in surface waters using LC/APCI/tandem mass spectrometry◊." *Analytical and bioanalytical chemistry* 384, no. 5: 1236-1245.

[19] Hernández-Rodríguez, D., J. E. Sánchez, M. G. Nieto, and F. J. Márquez-Rocha. 2006. "Degradation of endosulfan during substrate preparation and cultivation of Pleurotus pulmonarius." *World Journal of Microbiology and Biotechnology* 22, no. 7: 753-760.

[20] Marzio, D., W., E. Sáenz, J. Alberdi, M. Tortorelli, P. Nannini, and G. Ambrini. 2005. "Bioaccumulation of endosulfan from contaminated sediment by Vallisneria spiralis." *Bulletin of environmental contamination and toxicology* 74, no. 4: 637-644.

[21] DeLorenzo, M. E., L. A. Taylor, S. A. Lund, P. L. Pennington, E. D. Strozier, and M. H. Fulton. 2002. "Toxicity and bioconcentration potential of the agricultural pesticide endosulfan in phytoplankton and zooplankton." *Archives of Environmental Contamination and Toxicology* 42, no. 2: 173-181.

[22] Ramaneswari, K., and L. M. Rao. 2000. "Bioconcentration of endosulfan and monocrotophos by Labeo rohita and Channa punctata." *Bulletin of environmental contamination and toxicology* 65, no. 5: 618-622.

[23] Kumari, Beena, Jagdeep Singh, Shashi Singh, and T. S. Kathpal. 2005. "Monitoring of butter and ghee (clarified butter fat) for pesticidal contamination from cotton belt of Haryana, India." *Environmental Monitoring and Assessment* 105, no. 1-3: 111-120.

[24] Sutherland, Tara D., Irene Horne, Michael J. Lacey, Rebecca L. Harcourt, Robyn J. Russell, and John G. Oakeshott. 2000. "Enrichment of an endosulfan-degrading mixed bacterial culture." *Applied and Environmental Microbiology* 66:2822-2828.

[25] Sutherland, T. D., I. Horne, R. L. Harcourt, R. J. Russell, and J. G. Oakeshott. 2002. "Isolation and characterization of a Mycobacterium strain that metabolizes the insecticide endosulfan." *Journal of Applied Microbiology* 93:380-389.

[26] Martens, Rainer. 1976. "Degradation of [8, 9,-14C] endosulfan by soil microorganisms." *Applied and Environmental Microbiology* 31:853-858.

[27] Kwon, Gi-Seok, Ho-Yong Sohn, Kee-Sun Shin, Eungbin Kim, and Bu-Il Seo. 2005. "Biodegradation of the organochlorine insecticide, endosulfan, and the toxic metabolite, endosulfan sulfate, by Klebsiella oxytoca KE-8." *Applied Microbiology and Biotechnology* 67:845-850.

[28] Mukherjee, Irani, and Madhuban Gopal. 1994. "Degradation of β--endosulfan by Aspergillus Niger." *Toxicological and Environmental Chemistry* 46:217-221.

[29] Kullman, Seth W., and Fumio Matsumura. 1996. "Metabolic pathways utilized by Phanerochaete chrysosporium for degradation of the cyclodiene pesticide endosulfan." *Applied and Environmental Microbiology* 62:593-600.

[30] Awasthi, N., A. K. Singh, R. K. Jain, B. S. Khangarot, and A. Kumar. 2003. "Degradation and detoxification of endosulfan isomers by a defined co-culture of two Bacillus strains." *Applied Microbiology and Biotechnology* 62:279-283.

[31] Awasthi, N., N. Manickam, and A. Kumar. 1997. "Biodegradation of endosulfan by a bacterial coculture." *Bulletin of Environmental Contamination and Toxicology* 59: 928-934.

[32] Siddique, Tariq, Benedict C. Okeke, Muhammad Arshad, and William T. Frankenberger. 2003b. "Biodegradation kinetics of endosulfan by *Fusarium ventricosum* and a *Pandoraea* species." *Journal of Agricultural and Food Chemistry* 51:8015-8019.

[33] Sharma, Priyanka, Santosh K. Sharma, Anil Sharma, Ashutosh Sharma, and Pradeep Paraher. 2013. "Biodegradation of endosulfan using microbial culture." *International Journal of Recent Research and Review* 6.

[34] Alexander, M. 1994. "Biodegradation and Bioremediation Academic Press Inc." *San Diego, California* 267:269.

[35] U.S. EPA, 2002. *Reregistration Eligibility Decision for Endosulfan.* In: Prevention, P.a.T. S., (Ed.) EPA 738-R-02-013.

[36] Jia, Hongliang, Liyan Liu, Yeqing Sun, Bing Sun, Degao Wang, Yushan Su, Kurunthachalam Kannan, and Yi-Fan Li. 2010. "Monitoring and modeling endosulfan in Chinese surface soil." *Environmental Science and Technology* 44:9279-9284.

[37] Chakraborty, Paromita, Gan Zhang, Jun Li, Yue Xu, Xiang Liu, Shinsuke Tanabe, and Kevin C. Jones. 2010. "Selected organochlorine pesticides in the atmosphere of major Indian cities: levels, regional versus local variations, and sources." *Environmental Science and Technology* 44:8038-8043.

[38] Singh, Vandana, and Nandita Singh. 2014. "Uptake and accumulation of endosulfan isomers and its metabolite endosulfan sulfate in naturally growing plants of contaminated area." *Ecotoxicology and Environmental Safety* 104:189-193.

[39] Goncalves, C., and M. F. Alpendurada. 2005. "Assessment of pesticide contamination in soil samples from an intensive horticulture area, using ultrasonic extraction and gas chromatography–mass spectrometry." *Talanta* 65:1179-1189.

[40] Kumari, Beena, V. K. Madan, and T. S. Kathpal. 2008. "Status of insecticide contamination of soil and water in Haryana, India." *Environmental Monitoring and Assessment* 136:239-244.

[41] Chakraborty, Paromita, Gan Zhang, Jun Li, A. Sivakumar, and Kevin C. Jones. 2015. "Occurrence and sources of selected organochlorine pesticides in the soil of seven major Indian cities: Assessment of air–soil exchange." *Environmental Pollution* 204:74-80.

[42] Chakraborty, Paromita, Sanjenbam Nirmala Khuman, Bhupander Kumar, and Daniel Snow. 2018. "Sources of Organochlorine Pesticidal Residues in the Paddy Fields Along the Ganga-Brahmaputra River Basin: Implications for Long-Range Atmospheric Transport." In *Environmental Pollution of Paddy Soils*, pp. 69-83. Springer

[43] Ramesh, Atmakuru, and Ambalatharasu Vijayalakshmi. 2002. "Environmental exposure to residues after aerial spraying of endosulfan: residues in cow milk, fish, water, soil and cashew leaf in Kasargode, Kerala, India." *Pest Management Science: formerly Pesticide Science* 58:1048-1054.

[44] Pokhrel, Balram, Ping Gong, Xiaoping Wang, Mengke Chen, Chuanfei Wang, and Shaopeng Gao. 2018. "Distribution, sources, and air–soil exchange of OCPs, PCBs, and PAHs in urban soils of Nepal." *Chemosphere* 200:532-541.

[45] Hussain, Sarfraz, Muhammad Arshad, Muhammad Saleem, and Azeem Khalid. 2007a. "Biodegradation of α-and β-endosulfan by soil bacteria." *Biodegradation* 18:731-740.

[46] Hussain, Sarfraz, Muhammad Arshad, Muhammad Saleem, and Zahir Ahmad Zahir. 2007b. "Screening of soil fungi for in vitro degradation of endosulfan." *World Journal of Microbiology and Biotechnology* 23:939-945.

[47] Arshad, M., S. Hussain, and M. Saleem. 2008. "Optimization of environmental parameters for biodegradation of α- and β-endosulfan in soil slurry by Pseudomonas aeruginosa." *Journal of Applied Microbiology* 104:364-370.

[48] Jiang, Yu-Feng, Xue-Tong Wang, Ying Jia, Fei Wang, Ming-Hong Wu, Guo-Ying Sheng, and Jia-Mo Fu. 2009. "Occurrence, distribution and possible sources of organochlorine pesticides in agricultural soil of Shanghai, China." *Journal of Hazardous Materials* 170: 989-997.

[49] Fang, Yanyan, Zhiqiang Nie, Qingqi Die, Yajun Tian, Feng Liu, Jie He, and Qifei Huang. 2016. "Spatial distribution, transport dynamics, and health risks of endosulfan at a contaminated site." *Environmental pollution* 216:538-547.

[50] Menone, Mirta L., Silvia F. Pesce, María P. Díaz, Víctor J. Moreno, and Daniel A. Wunderlin. 2008. "Endosulfan induces oxidative stress and changes on detoxication enzymes in the aquatic macrophyte *Myriophyllum quitense.*" *Phytochemistry* 69:1150-1157.

[51] Fang, Yanyan, Zhiqiang Nie, Jinzhong Yang, Qingqi Die, Yajun Tian, Feng Liu, Jie He, Jianyuan Wang, and Qifei Huang. 2018. "Spatial distribution of and seasonal variations in endosulfan concentrations in soil, air, and biota around a contaminated site." *Ecotoxicology and Environmental Safety* 161:402-408.

[52] Raju, Jaishree, and V. K. Gupta. 1991. "A simple spectrophotometric determination of endosulfan in river water and soil." *Fresenius' Journal of Analytical Chemistry* 339:431-433.

[53] Pasha, Chand, and Badiadka Narayana. 2008. "Spectrophotometric determination of endosulfan using thionin and methylene blue as chromogenic reagents." *Bulletin of Environmental Contamination and Toxicology* 80:85-89.

[54] Venugopal, N. V. S., and B. Sumalatha. 2011. "Spectrophotometric determination of endosulfan in environmental samples." In *2nd international conference on environmental science and technology*, vol. 6, pp. 195-197.

[55] London L. 1992. An overview of agrochemical hazards in the South African farming sector. *S Afr Med J.* 81:560-564.

[56] Fantke, Peter, Assumpcio Anton, Tim Grant, and Kiyotada Hayashi. 2017. "Pesticide emission quantification for life cycle assessment: A global consensus building process." *International Journal of Life Cycle Assessment* 13:245-251.

[57] Sankararamakrishnan, Nalini, Ajit Kumar Sharma, and Rashmi Sanghi. 2005. "Organochlorine and organophosphorous pesticide residues in ground water and surface waters of Kanpur, Uttar Pradesh, India." *Environment International* 31:113-120.

[58] Miliadis, George Emm. 1994. "Determination of pesticide residues in natural waters of Greece by solid phase extraction and gas chromatography." *Bulletin of Environmental Contamination and Toxicology* 52:25-30.

[59] Sun, Runxia, Xiaojun Luo, Bin Tang, Zongrui Li, Tao Wang, Lin Tao, and Bixian Mai. 2016. "Persistent halogenated compounds in fish from rivers in the Pearl River Delta, South China: Geographical pattern and implications for anthropogenic effects on the environment." *Environmental Research* 146:371-378.

[60] Pan, Hongwei, Hongjun Lei, Xiaosong He, Beidou Xi, and Qigong Xu. 2017. "Spatial distribution of organochlorine and organophosphorus pesticides in soil-groundwater systems and their associated risks in the middle reaches of the Yangtze River Basin." *Environmental Geochemistry and Health* 1-13.

[61] Tuxen, Nina, Peter L. Tüchsen, Kirsten Rügge, Hans-Jørgen Albrechtsen, and Poul L. Bjerg. 2000. "Fate of seven pesticides in an aerobic aquifer studied in column experiments." *Chemosphere* 41:1485-1494.

[62] Dalvie, Mohamed A., Eugene Cairncross, Abdullah Solomon, and Leslie London. 2003. "Contamination of rural surface and ground water by endosulfan in farming areas of the Western Cape, South Africa." *Environmental Health* 2: 1.

[63] Chaza, Chbib, Net Sopheak, Hamzeh Mariam, Dumoulin David, Ouddane Baghdad, and Baroudi Moomen. 2018. "Assessment of pesticide contamination in Akkar groundwater, northern Lebanon." *Environmental Science and Pollution Research* 25:14302-14312.

[64] Navarrete, Ian A., Kendric Aaron M. Tee, Jewel Racquel S. Unson, and Arnold V. Hallare. 2018. "Organochlorine pesticide residues in surface water and groundwater along Pampanga River, Philippines." *Environmental Monitoring and Assessment* 190: 289.

[65] Bai, Ying, Xiaohong Ruan, and J. P. van der Hoek. 2018. "Residues of organochlorine pesticides (OCPs) in aquatic environment and risk assessment along Shaying River, China." *Environmental Geochemistry and Health* 1-14.

[66] Ozmen, O., and F. Mor. 2015. "Effects of vitamin C on pathology and caspase-3 activity of kidneys with subacute endosulfan toxicity." *Biotechnic and Histochemistry* 90:25-30.

[67] Choudhary, N., and S. C. Joshi. 2003. "Reproductive toxicity of endosulfan in male albino rats." *Bulletin of Environmental Contamination and Toxicology* 70:0285-0289.

[68] Jamil, Kaiser, Abjal Pasha Shaik, M. Mahboob, and D. Krishna. 2004. "Effect of organophosphorus and organochlorine pesticides (monochrotophos, chlorpyriphos, dimethoate, and endosulfan) on human lymphocytes *in-vitro*." *Drug and Chemical Toxicology* 27:133-144.

[69] Aggarwal, Manoj, Suresh Babu Naraharisetti, S. Dandapat, G. H. Degen, and J. K. Malik 2008. "Perturbations in immune responses induced by concurrent subchronic exposure to arsenic and endosulfan." *Toxicology* 251:51-60.

[70] Terry, Alexander I., Sandra Benitez-Kruidenier, and Gregory K. DeKrey. 2018. "Effects of endosulfan isomers on cytokine and nitric oxide production by differentially activated RAW 264.7 cells." *Toxicology Reports* 5:396-400.

[71] Ghosh, Krishna, Biji Chatterjee, Aparna Geetha Jayaprasad, and Santosh R. Kanade. 2018. "The persistent organochlorine pesticide endosulfan modulates multiple epigenetic regulators with oncogenic potential in MCF-7 cells." *Science of the Total Environment* 624:1612-1622.

[72] Zhang, Lianshuang, Jialiu Wei, Lihua Ren, Jin Zhang, Man Yang, Li Jing, Ji Wang, Zhiwei Sun, and Xianqing Zhou. 2017. "Endosulfan inducing apoptosis and necroptosis through activation RIPK signaling pathway in human umbilical vascular endothelial cells." *Environmental Science and Pollution Research* 24:215-225.

[73] Sebastian, R., and S. C. Raghavan. 2015. "Endosulfan induces male infertility." e2022.

[74] Wei, Jialiu, Lianshuang Zhang, Lihua Ren, Jin Zhang, Yang Yu, Ji Wang, Junchao Duan, Cheng Peng, Zhiwei Sun, and Xianqing Zhou. 2017. "Endosulfan inhibits proliferation through the Notch signaling pathway in human umbilical vein endothelial cells." *Environmental Pollution* 221:26-36.

[75] Rau, A. T. K., Anita Coutinho, K. Shreedhara Avabratha, Aarathi R. Rau, and Raj P. Warrier. 2012. "Pesticide (endosulfan) levels in the bone marrow of children with hematological malignancies." *Indian Pediatrics* 49:113-117.

[76] Xu, Dan, Dong Liang, Yubing Guo, and Yeqing Sun. 2018. "Endosulfan causes the alterations of DNA damage response through ATM-p53 signaling pathway in human leukemia cells." *Environmental Pollution* 238:1048-1055.

[77] Somtrakoon, Khanitta, and Maleeya Kruatrachue. 2013. "Influence of indolebutyric acid on Brassica chinensis seedling growth growing in endosulfan-sulfate contaminated sand." *Journal of Agricultural Research and Extension* 30:14-24.

[78] Singh, Amit Kishore, Prem Pratap Singh, Vijay Tripathi, Hariom Verma, Sandeep Kumar Singh, Akhileshwar Kumar Srivastava, and Ajay Kumar. 2018. "Distribution of cyanobacteria and their interactions with pesticides in paddy field: A comprehensive review." *Journal of Environmental Management* 224:361-375.

[79] Igbedioh, S. O. 1991. "Effects of agricultural pesticides on humans, animals, and higher plants in developing countries." *Archives of Environmental Health: An International Journal* 46: 218-224.

[80] Akobundu, I. Okezie. 1987. Safe use of herbicides. *Weed Science in the Tropics. Principles and Practices. John Wiley & Sons, New York*, pp. 318-334.

[81] Waseem, Raja, and Rathaur Preeti. 2012. "Endosulfan induced changes in growth rate, pigment composition and photosynthetic activity of mosquito fern Azolla microphylla." *Journal of Stress Physiology and Biochemistry* 8, 4.

[82] Mathad, Pratima, and N. C. Siddaling. 2009. "Antioxidant status of pigeon pea, Cajanus cajan in the presence of endosulfan stress." *Journal of Environmental Biology* 30:451-454.

[83] Kumar, Satyendra, Khalid Habib, and Tasneem Fatma. 2008. "Endosulfan induced biochemical changes in nitrogen-fixing cyanobacteria." *Science of the Total Environment* 403:130-138.

[84] Prasad, Sheo Mohan, Deelip Kumar, and Mohd Zeeshan. 2005. "Growth, photosynthesis, active oxygen species and antioxidants responses of paddy field cyanobacterium Plectonema boryanum to endosulfan stress." *The Journal of General and Applied Microbiology* 51: 115-123.

[85] Prasad, Sheo Mohan, Mohd Zeeshan, and Deelip Kumar. 2011. "Toxicity of endosulfan on growth, photosynthesis, and nitrogenase activity in two species of Nostoc (Nostoc muscorum and Nostoc calcicola)." *Toxicological and Environmental Chemistry* 93: 513-525.

[86] Rani, Rupa, Vipin Kumar, Pratishtha Gupta, and Avantika Chandra. 2019. "Effect of endosulfan tolerant bacterial isolates (Delftia lacustris IITISM30 and Klebsiella aerogenes IITISM42) with Helianthus annuus on remediation of endosulfan from contaminated soil." *Ecotoxicology and environmental safety* 168:315-323.

[87] Pérez, Débora J., Mirta L. Menone, Elsa L. Camadro, and Víctor J. Moreno. 2008. "Genotoxicity evaluation of the insecticide endosulfan in the wetland macrophyte Bidens laevis L." *Environmental Pollution* 153:695-698.

[88] Chouychai, Waraporn. 2012. "Effect of some plant growth regulators on lindane and α--endosulfan toxicity to *Brassica chinensis*." *Journal of Environmental Biology* 33:811.

[89] Suri, K. S., and Gursharan Singh. 2011. "Insecticide-induced resurgence of the whitebacked planthopper Sogatella furcifera (Horvath) (Hemiptera: Delphacidae) on rice varieties with different levels of resistance." *Crop Protection* 30:118-124.

[90] Dhungana, Sanjeev Kumar, Il-Doo Kim, Hwa-Sook Kwak, and Dong-Hyun Shin. 2016. "Unraveling the effect of structurally different classes of insecticide on germination and early plant growth of soybean [Glycine max (L.) Merr.]." *Pesticide Biochemistry and Physiology* 130:39-43.

[91] Lee, Nanju, John H. Skerritt, and David P. McAdam. 1995. "Hapten synthesis and development of ELISAs for detection of endosulfan in water and soil." *Journal of Agricultural and Food Chemistry* 43:1730-1739.

[92] Iqbal, Z., A. Hussain, A. Latif, M. R. Asi, and J. A. Chaudhary. 2001. "Impact of pesticide applications in cotton agro ecosystem and soil bioactivity studies I: microbial populations." *Journal of Biological Sciences* 7:640-644.

[93] Defo, Michel Amery, Thomas Njine, Moise Nola, and Florence Sidonie Beboua. 2011. "Microcosm study of the long term effect of endosulfan on enzyme and microbial activities on two agricultural soils of Yaounde-Cameroon." *African Journal of Agricultural Research* 6:2039-205.

[94] Bloomquist, Jeffrey R. 1996. "Ion channels as targets for insecticides." *Annual Review of Entomology* 41:163-190.

[95] Khan, Jamal A., Samiullah Khan, and Sarwat F. Usmani. 2013. "Effect of Endosulfan on seed germination, growth and nutrients uptake of fenugreek plant." *Journal of Industrial Research & Technology* 2:88-91.

[96] Shanmugam, M., M. Venkateshwarlu, and A. Naveed. 2000. "Effect of pesticides on the freshwater crab Barytelphusa cunicularis (Westwood)." *Journal of Ecotoxicology and Environmental Monitoring* 10:273-279.

[97] Chaudhuri, K., S. Selvaraj, and A. K. Pal. 1999. "Studies on the genotoxicity of endosulfan in bacterial systems." *Mutation Research/Genetic Toxicology and Environmental Mutagenesis*439, no. 1: 63-67.

[98] ASTDR. 2000. *Toxicological profile for endosulfan US Department of health and human services*; Aegncy for toxic substances and disease registry, Atlanta, Ga, USA.

[99] Lambert, Michael RK. 1997. "Effects of pesticides on amphibians and reptiles in sub-Saharan Africa." In *Reviews of environmental contamination and toxicology*, pp. 31-73. Springer, New York, NY.

[100] Yonli, Arsène H., Isabelle Batonneau-Gener, and Jean Koulidiati. 2012. "Adsorptive removal of α-endosulfan from water by hydrophobic zeolites. An isothermal study." *Journal of Hazardous Materials* 203:357-362.

[101] Pignatello, Joseph J., Esther Oliveros, and Allison MacKay. 2006. "Advanced oxidation processes for organic contaminant destruction based on the Fenton reaction and related chemistry." *Critical reviews in environmental science and technology* 36, no. 1:1-84.

[102] Singh, Swatantra P., and Purnendu Bose. 2017. "Reductive dechlorination of endosulfan isomers and its metabolites by zero-valent metals: reaction mechanism and degradation products." *RSC Advances* 7: 27668-27677.

[103] Begum, Asfiya, and Sumit Kumar Gautam. 2012. "Endosulfan and lindane degradation using ozonation." *Environmental Technology* 33:943-949.

[104] Cong, Lujing, Jing Guo, Jisong Liu, Haiyan Shi, and Minghua Wang. 2015. "Rapid degradation of endosulfan by zero-valent zinc in water and soil." *Journal of Environmental Management* 150:451-455.

[105] Gupta, Vinod K., and Imran Ali. 2008. "Removal of endosulfan and methoxychlor from water on carbon slurry." *Environmental Science and Technology* 42:766-770.

[106] Peñuela, Gustavo A., and Damià Barceló. 1998. "Application of C18 disks followed by gas chromatography techniques to degradation kinetics, stability and monitoring of endosulfan in water." *Journal of Chromatography A* 795:93-104.

[107] Shah, Noor S., Xuexiang He, Javed Ali Khan, Hasan M. Khan, Dominic L. Boccelli, and Dionysios D. Dionysiou. 2015. "Comparative studies of various iron-mediated oxidative systems for the photochemical degradation of endosulfan in aqueous solution." *Journal of Photochemistry and Photobiology A: Chemistry* 306:80-86.

[108] Xiong, Bailian, Anhong Zhou, Guocan Zheng, Jinzhong Zhang, and Weihong Xu. 2015. "Photocatalytic degradation of endosulfan in contaminated soil with the elution of surfactants." *Journal of Soils and Sediments* 15:1909-1918.

[109] Guerin, Turlough F. 1999. "The anaerobic degradation of endosulfan by indigenous microorganisms from low-oxygen soils and sediments." *Environmental Pollution* 106: 13-21.

[110] Bhalerao, Tejomyee S., and Pravin R. Puranik. 2007. "Biodegradation of organochlorine pesticide, endosulfan, by a fungal soil isolate, Aspergillus niger." *International Biodeterioration and Biodegradation* 59:315-321.

[111] Miles, J. R. W., and P. Moy. 1979. "Degradation of endosulfan and its metabolites by a mixed culture of soil microorganisms." *Bulletin of Environmental Contamination and Toxicology* 23:13-19.

[112] Li, Wen, Yun Dai, Beibei Xue, Yingying Li, Xiang Peng, Jingshun Zhang, and Yanchun Yan. 2009. "Biodegradation and detoxification of endosulfan in aqueous medium and soil by *Achromobacter xylosoxidans* strain CS5." *Journal of Hazardous Materials* 167:209-216.

[113] Kumar, Koel, Sivanesan Saravana Devi, Kannan Krishnamurthi, Gajanan Sitaramji Kanade, and Tapan Chakrabarti. 2007. "Enrichment and isolation of endosulfan degrading and detoxifying bacteria." *Chemosphere* 68, no. 2: 317-322.

[114] Javaid, Muhammad Kashif, Mehrban Ashiq, and Muhammad Tahir. 2016. "Potential of biological agents in decontamination of agricultural soil." *Scientifica*.

[115] Kumar M and Philip L. 2006. Endosulfan mineralization by bacterial isolates and possible degradation pathway identification. *Bioremediation J 10:179–190*

[116] Ozdal, Murat, Ozlem Gur Ozdal, Omer Faruk Algur, and Esabi Basaran Kurbanoglu. 2017. "Biodegradation of α-endosulfan via hydrolysis pathway by *Stenotrophomonas maltophilia* OG2." *3 Biotech* 7: 113.

[117] Katayama, Arata, and Fumio Matsumura. 1993. "Degradation of organochlorine pesticides, particularly endosulfan, by *Trichoderma harzianum*." *Environmental Toxicology and Chemistry: An International Journal* 12:1059-1065.

[118] Cáceres, Tanya P., Mallavarapu Megharaj, and Ravi Naidu. 2008. "Biodegradation of the pesticide fenamiphos by ten different species of green algae and cyanobacteria." *Current microbiology* 57:643-646.

[119] Lee, Sung-Eun, Jong-Soo Kim, Ivan R. Kennedy, Jong-Woo Park, Gi-Seok Kwon, Sung-Cheol Koh, and Jang-Eok Kim. 2003. "Biotransformation of an organochlorine insecticide, endosulfan, by *Anabaena* species." *Journal of Agricultural and Food Chemistry* 51:1336-1340.

[120] Weir, Kahli M., Tara D. Sutherland, Irene Horne, Robyn J. Russell, and John G. Oakeshott. 2006. "A single monooxygenase, ese, is involved in the metabolism of the organochlorides endosulfan and endosulfate in an Arthrobacter sp." *Applied and Environmental Microbiology* 72:3524-3530.

[121] Schrijver, Adinda De, and René De Mot. 1999. "Degradation of pesticides by actinomycetes." *Critical Reviews in Microbiology* 25:85-119.

[122] Briceño, Gabriela, Graciela Palma, and Nelson Durán. 2007. "Influence of organic amendment on the biodegradation and movement of pesticides." *Critical Reviews in Environmental Science and Technology* 37:233-271.

[123] Diez, M. C. 2010. "Biological aspects involved in the degradation of organic pollutants." *Journal of Soil Science and Plant Nutrition* 10:244-267.

[124] Lal, Rup, and D. M. Saxena. 1982. "Accumulation, metabolism, and effects of organochlorine insecticides on microorganisms." *Microbiological Reviews* 46:95.

[125] Widada, J., H. Nojiri, and T. Omori. 2002. "Recent developments in molecular techniques for identification and monitoring of xenobiotic-degrading bacteria and their catabolic genes in bioremediation." *Applied Microbiology and Biotechnology* 60: 45-59.

[126] Ortiz-Hernández, Ma Laura, Enrique Sánchez-Salinas, Edgar Dantán-González, and María Luisa Castrejón-Godínez. 2013. "Pesticide biodegradation: mechanisms, genetics and strategies to enhance the process." In *Biodegradation-life of science*. Intech.

[127] Singh, Brajesh K., and Allan Walker. 2006. "Microbial degradation of organophosphorus compounds." *FEMS Microbiology Reviews* 30:428-471.

[128] Sharma, Anita, Pankaj, Priyanka Khati, Saurabh Gangola, and Govind Kumar. 2016. "Microbial degradation of pesticides for environmental cleanup." *Bioremediation of Industrial Pollutants*.

[129] Sims, Gerald K., and Alison M. Cupples. 1999. "Factors controlling degradation of pesticides in soil." *Pesticide Science* 55:598-601.

[130] Awasthi, Niranjan, Rajiv Ahuja, and Ashwani Kumar. 2000. "Factors influencing the degradation of soil-applied endosulfan isomers." *Soil Biology and Biochemistry* 32: 1697-1705.

[131] Castillo, Maria del Pilar, and Lennart Torstensson. 2007. "Effect of biobed composition, moisture, and temperature on the degradation of pesticides." *Journal of Agricultural and Food Chemistry* 55:5725-5733.

[132] Lee, Sung-Eun, Jong-Soo Kim, Ivan R. Kennedy, Jong-Woo Park, Gi-Seok Kwon, Sung-Cheol Koh, and Jang-Eok Kim. 2003. "Biotransformation of an organochlorine insecticide, endosulfan, by Anabaena species." *Journal of Agricultural and Food Chemistry* 51:1336-1340.

[133] Reddy, P. Manoj Kumar, Sk Mahammadunnisa, and Ch Subrahmanyam. 2014. "Catalytic non-thermal plasma reactor for mineralization of endosulfan in aqueous medium: A green approach for the treatment of pesticide contaminated water." *Chemical Engineering Journal* 238:157-163.

[134] Banasiak, Laura J., Bart Van der Bruggen, and Andrea I. Schäfer. 2011. "Sorption of pesticide endosulfan by electrodialysis membranes." *Chemical Engineering Journal* 166, no. 1: 233-239.

[135] Jayashree, R., N. Vasudevan, and S. Chandrasekaran. 2006. "Surfactants enhanced recovery of endosulfan from contaminated soils." *International Journal of Environmental Science and Technology* 3:251-259.

[136] Shah, Noor S., Xuexiang He, Hasan M. Khan, Javed Ali Khan, Kevin E. O'Shea, Dominic L. Boccelli, and Dionysios D. Dionysiou. 2013. "Efficient removal of endosulfan from aqueous solution by UV-C/peroxides: a comparative study." *Journal of Hazardous Materials* 263: 584-592.

[137] Thomas, Jesty, K. Praveen Kumar, and K. R. Chitra. 2011. "Synthesis of Ag doped nano TiO_2 as efficient solar photocatalyst for the degradation of endosulfan." *Advanced Science Letters* 4:108-114.

[138] Kida, Małgorzata, Sabina Ziembowicz, and Piotr Koszelnik. 2018. "Removal of organochlorine pesticides (OCPs) from aqueous solutions using hydrogen peroxide, ultrasonic waves, and a hybrid process." *Separation and Purification Technology* 192:457-464.

[139] Sivagami, K., B. Vikraman, R. Ravi Krishna, and T. Swaminathan. 2016. "Chlorpyrifos and Endosulfan degradation studies in an annular slurry photo reactor." *Ecotoxicology and environmental safety* 134:327-331.

[140] Sarangapani, C., Misra, N.N., Milosavljevic, V., Bourke, P., O'Regan, F. and Cullen, P.J., 2016. "Pesticide degradation in water using atmospheric air cold plasma". *Journal of Water Process Engineering* 9:225-232.

[141] Singh, Swatantra Pratap, and Purnendu Bose. 2016. "Degradation kinetics of Endosulfan isomers by micron-and nano-sized zero valent iron particles (MZVI and NZVI)." *Journal of Chemical Technology & Biotechnology* 91:2313-2321.

[142] Begum, Asfiya, Prakhar Agnihotri, Amit B. Mahindrakar, and Sumit Kumar Gautam. 2017. "Degradation of endosulfan and lindane using Fenton's reagent." *Applied Water Science* 7: 207-215.

[143] Kafilzadeh, Farshid, Moslem Ebrahimnezhad, and Yaghoob Tahery. 2015. "Isolation and identification of endosulfan-degrading bacteria and evaluation of their bioremediation in Kor River, Iran." *Osong Public Health and Research Perspectives* 6:39-46.

[144] Guha, A., B. Kumari, T. C. Bora, P. C. Deka, and M. K. Roy. 2000. "Bioremediation of endosulfan by Micrococcus sp." *Indian Journal of Environmental Health* 42:9-12.

[145] Kumar, Mohit, C. Vidya Lakshmi, and Sunil Khanna. 2008. "Biodegradation and bioremediation of endosulfan contaminated soil." *Bioresource Technology* 99:3116-3122.

[146] Goswami, Supriya, and Dileep K. Singh. 2009. "Biodegradation of α and β endosulfan in broth medium and soil microcosm by bacterial strain Bordetella sp. B9." *Biodegradation* 20:199-207.

[147] Kong, Lingfen, Yu Zhang, Lusheng Zhu, Jinhua Wang, Jun Wang, Zhongkun Du, and Cheng Zhang. 2018. "Influence of isolated bacterial strains on the in situ biodegradation of endosulfan and the reduction of endosulfan-contaminated soil toxicity." *Ecotoxicology and Environmental Safety* 160:75-83.

[148] Zaffar, Habiba, Raza Ahmad, Arshid Pervez, and Tatheer Alam Naqvi. 2018. "A newly isolated *Pseudomonas* sp. can degrade endosulfan via hydrolytic pathway." *Pesticide Biochemistry and Physiology* 152:69-75.

[149] Shetty, P. K., Jharna Mitra, N. B. K. Murthy, K. K. Namitha, K. N. Savitha, and K. Raghu. 2000. "Biodegradation of cyclodiene insecticide endosulfan by Mucor thermohyalospora MTCC 1384." *Current Science* 1381-1383.

[150] Mukherjee, I., and A. Mittal. 2005. "Bioremediation of endosulfan using *Aspergillus terreus* and *Cladosporium oxysporum*." *Bulletin of Environmental Contamination and Toxicology* 75: 1034-1040.

[151] Gangola, Saurabh, Geeta Negi Pankaj, Anjana Srivastava, and Anita Sharma. 2015. "Enhanced Biodegradation of Endosulfan by *Aspergillus* and *Trichoderma* spp. Isolated from an Agricultural Field of Tarai Region of Uttarakhand." *Pesticide Research Journal* 27:223-230.

[152] Abraham, Jayanthi, and Sivagnanam Silambarasan. 2014. "Biomineralization and formulation of endosulfan degrading bacterial and fungal consortiums". *Pesticide Biochemistry and Physiology* 116:24-31.

[153] Mohapatra, P. K., and R. C. Mohanty. 1992. "Growth pattern changes of Chlorella vulgaris and Anabaena doliolum due to toxicity of dimethoate and endosulfan." *Bulletin of Environmental Contamination and Toxicology* 49:576-581.

[154] Jha, M. N., and S. K. Mishra. 2005. "Biological responses of cyanobacteria to insecticides and their insecticide degrading potential." *Bulletin of Environmental Contamination and Toxicology* 75:374-381.

Chapter 8

BIODEGRADATION AND DETOXIFICATION OF CHLORINATED PESTICIDE ENDOSULFAN BY SOIL MICROBES

Arpan Mukherjee[1], Vipin Kumar Singh[2] and Somenath Das[2,]*

[1]Institute of Environment and Sustainable Development, Banaras Hindu University, Varanasi
[2]Department of Botany, Centre of Advanced Study, Institute of Science, Banaras Hindu University, Varanasi

ABSTRACT

Endosulfan is a chlorinated cyclodiene insecticide which significantly impacts human health as well as the natural environment. The physicochemical methods based remediation techniques for endosulfan contaminated sites are costly and produce a large amount of secondary sludge of even more hazardous nature in some cases as compared to the parent compound. Application of microbes for the management of

---

*Corresponding Author's Email: sndbhu@gmail.com.

endosulfan is a nature-friendly, less expensive and sustainable approach to mitigate the pollutant toxicity. Numerous microbial species has been investigated *in vitro* to explore the feasibility of indigenous strains for contaminant removal. However, a field-scale application requires the detailed investigations of their mechanism of action and functionality under natural environmental conditions. Selection and application of efficient endosulfan degrading microbial species would be imperative for remediation practices. In this chapter, we have discussed different types of microbes (bacteria/cyanobacteria, fungi, and algae) involved in the degradation of the endosulfan in the soil.

Keywords: Endosulfan, endosulfan sulfate, monooxygenase, neurotoxic, pesticide, biodegradation

1. INTRODUCTION

Endosulfan ($C_9H_6C_{16}O_3S$) is an extensively used synthetic chlorinated cyclodiene insecticide which has been widely reported to be present in air, surface soil, sediments, and in rain waters [1]. It was first commercially utilized in 1954 [2]. Among South Asian countries, India is one of the largest consumers of endosulfan utilizing extensively for the cultivation of cotton crops [3]. According to the recent report of Indian Chemical Council (ICC), endosulfan is extensively used to control pests in a wide range of plants and crops such as tea, coffee, cotton, cereals, fruits, cashew, and vegetables. It is chemically synthesized by mixing two isomers i.e., α-endosulfan and β-endosulfan in the proportion of 70:30, respectively. Indiscriminate utilization of endosulfan has resulted in the contamination of soil through its non-biodegradable residues. Long term existence of endosulfan depends on the half-life of its sulfur derivatives, for example, 60 days for α-endosulfan and 800 days for β-endosulfan. The lower degradation rate of endosulfan together with the formation of endosulfan sulfate, an intermediate metabolite may likely increase its toxicity in humans [4]. It can easily disrupt the endocrine systems of animals and acts as a potent genotoxic agent [5]. Furthermore, endosulfan may also affect the central nervous system (CNS), kidney, liver, blood, parathyroid gland, reproductive system and has mutagenic effects on animals [6]. Endosulfan has shown extreme toxicity to aquatic flora and fauna with its concentration ranging from 0.10-166 µg/L [7]. Because of persistent nature, endosulfan presence in the environment has emerged as a challenging issue, calling for its degradation or conversion through microbial activities into non-toxic or less hazardous forms. Since the chemical treatment technologies are expensive and produce a large number of secondary by-products, endosulfan degradation by deploying microbes is a viable and promising option. A number of microbes are reported to proliferate efficiently in endosulfan contaminated soil and can utilize the endosulfan and their metabolites as their energy source. Biological degradation and detoxification of endosulfan through microbes has proven to be a significant alternative against conventional treatment techniques such as landfill and incineration. Different groups of bacteria *viz. Pseudomonas aeruginosa*, *P. spinosa*, *Arthobacter* spp., *Mycobacterium* spp., *Burkholdaria cepacia*, fungi such as *Trichoderma harzinianum*, *Phanerochaete chrysosporium*, *Aspergillus terreus*, *A. sydowii*, *Cladosporium oxysporium* and microalgal or cyanobacterial strains *viz. Chlorococcum*, *Scenedesmus*, *Nostoc*, *Aulosira*, and *Anabaena* have been successfully utilized for the degradation of endosulfan under field conditions.

2. Effect of Endosulfan on Health

Since no effective alternative to endosulfan has been developed; this hazardous pesticide is still being used extensively in India and China as insecticide and acaricide. Due to its potential toxicities to a number of organisms including a mammal, its broad spectrum utilization has been banned in some countries [8]. It acts as a neurotoxin and inhibits the GABA (gamma amino butyric acid) gated chloride channels [9]. The major toxic effects of endosulfan include hyperstimulation of central nervous system (CNS), systemic organ deficiency, reproductive damage, immunodeficiency, genotoxicity, and endocrine system disruption in human beings [10, 11]. The endosulfan isomers have been represented to express differential toxicity. The β-endosulfan is comparatively more neurotoxic in humans, especially for human nervous SH-SY5Y and hepatic HepG2 cells [12, 13] than α-endosulfan. Deposition of endosulfan in the human body can disrupt the full endocrine system [14] by altering the levels of steroid hormones. It can also disturb the function of the testis and cause sperm abnormalities in mammals [15, 16]. An epidemiological study on village children in Northern India has shown that the exposure of endosulfan by aerial spraying can disturb the synthesis of sex hormone and can cause a delay in sexual maturity [17]. An illustrated account of endosulfan fate, transformation, and effects on human health is presented in Figure 1.

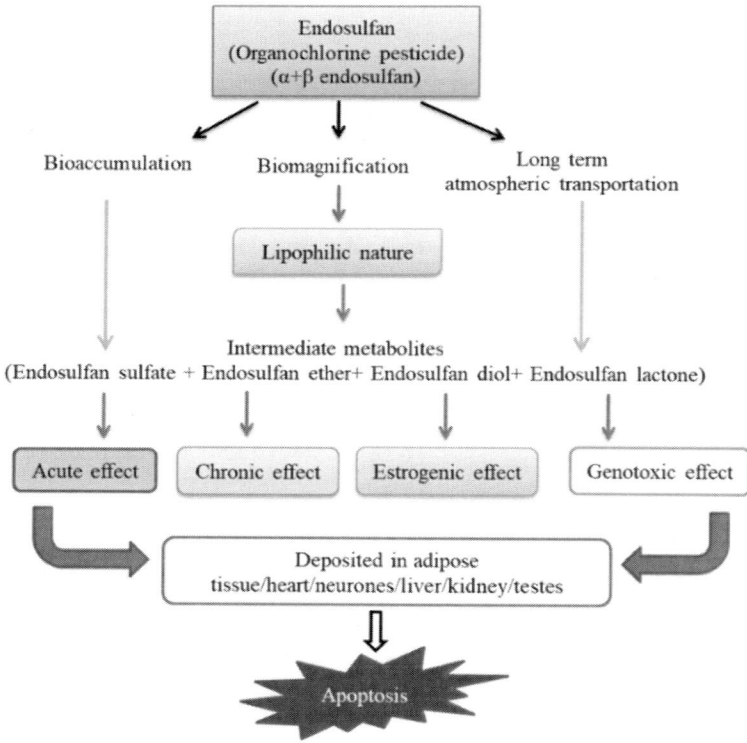

Figure 1. Flow chart showing an illustrated account of endosulfan fate, transformation and effects on human health.

3. MICROBIAL DEGRADATION OF ENDOSULFAN

A number of physicochemical and biological remedial techniques have been employed for the degradation and detoxification of endosulfan [18-22]. In biological remediation technique, different groups of soil microbes can utilize the endosulfan as a major source of carbon and sulfur during aerobic and anaerobic exposure to the substrate. Therefore, microbial degradation is a cost-effective and eco-friendly alternative to solve the endosulfan enrichment problem in different components of an ecosystem. Siddique et al. and Sachdev et al. [23, 24] have reported that a mixed culture of different bacterial species (consortia) may serve as an efficient and promising tool to degrade the endosulfan as compared to fungal cultures. Furthermore, the anaerobic biodegradation of endosulfan by microbes may be quite efficient as compared to aerobic degradation process in some cases.

3.1. Endosulfan Degradation by Bacteria

Among different bioremediation techniques, the bacterial degradation of endosulfan is much popular. Biologically produced surfactants may reduce soil water interfacial tension and thus may favor the enhanced degradation [24]. Jayashree and Vasudevan [25] reported degradation up to 94% and 87% of total endosulfan by *Pseudomonas aeruginosa* within 56-60 days. The major problem for endosulfan biodegradation is accumulation of endosulfan sulfate, an equally toxic metabolic intermediate [26]. Therefore, non-oxidative endosulfan degradation could be considered as a valid process for reduction of endosulfan sulfate. Very limited numbers of bacteria *viz. Pseudomonas aeruginosa* KE-1, KE-2 have been reported for degradation of endosulfan without formation of an intermediate compound i.e., endosulfan sulfate. However, little information is available regarding the factors influencing the dechlorination of endosulfan. Table 1 represents a brief account of endosulfan degrading bacterial species. Some of the important experimental investigations pertaining to endosulfan degradation by bacteria with their growth kinetics, metabolic intermediates, and detoxification rate are discussed briefly one by one.

In a study conducted by Singh and Singh [27], it was observed that *Achromobacter xylosoxidans* strain C8B can degrade endosulfan without forming any metabolites, and most importantly, the microbe utilized the endosulfan as the source of carbon and sulfur. The degradation pattern of endosulfan, its isomers and metabolic intermediates were detected through Gas chromatography-Electron capture detector (GC-ECD), High-Performance Liquid Chromatography (HPLC) and Thin Layer Chromatography (TLC). They concluded that the detoxification pattern of endosulfan was affected by the pH of the growth medium with a maximum pH of 6.8 for optimum growth. Further, the optimum temperature ranging from 28-35°C was found to be an important parameter for logarithmic growth of *Achromobacter xylosoxidans* to support the significant biodegradation of endosulfan.

Different strains of *Agrobacterium* are excellent degrader of chlorinated hydrocarbons and xenobiotic compounds in water and soil. *In vitro* degradation of endosulfan through *Agrobacterium tumefaciens* PT-3 has been reported within 3-4 days. Endosulfan at a concentration equivalent to 100 mg/L was almost completely utilized by the bacterium. The strains PT-3 was able to utilize both isomeric forms of endosulfan with equal efficiencies without accumulation of hazardous metabolic by-product. Biodegradation of endosulfan was affected by specific growth rate (μ), pH and incubation temperature of cultured media inoculated with *A. tumefaciens* EBRI25 and *A. tumefaciens* PT-3 strain having varied endosulfan concentrations [28]. The *Agrobacterium tumefaciens* may be suitably applied for bioaugmentation of endosulfan enriched soil to detoxify and ameliorate its toxicity on nitrogen-fixing bacterial species present in the soil ecosystem.

Sutherland et al. [29] and Verma et al. [30] reported that the strains of *Alkaligens faecalis* can grow abundantly in soil contaminated with endosulfan and can release specific enzymes for degradation of toxic pesticides. The supply of additional peptone into the medium served as the source of energy while sulfur for metabolic activities was acquired from endosulfan. The alkaline condition was observed to stimulate the growth of *Alkaligens faecalis* JBW$_4$. Another important point for the practical implication of *Alkaligens faecalis* JBW$_4$ was that it utilizes the hydrolytic pathway of endosulfan degradation, hence, reducing the formation of toxic intermediate compounds (endosulfan sulfate) [31].

Seralathan et al. [32] reported CYP-BM3 (Cytochrome P450 BM3) mediated effective bioremediation of endosulfan from soil and water. They firstly employed *in silico* approach for characterization of the gene and their products participating in endosulfan degradation. Abraham and Silambarasam [33] isolated *Klebsiella pneumoniae* JAS8 from rhizospheric soil and demonstrated its potency for biodegradation of α- and β-endosulfan; degradation of endosulfan sulfate followed first-order reaction kinetics within 10 days of incubation periods.

Kwon et al. [34] reported *Klebsiella oxytosa* KE-8 as a potential biocatalyst for active remediation of endosulfan as well as endosulfan sulfate from endosulfan-contaminated soil. The pH of the cultured cells decreased from 7.2 to 5.6 depending upon cell density and growth kinetics. Increase in biomass of *Klebsiella oxytosa* KE-8 treated with endosulfan-contaminated soils actively decreased endosulfan and endosulfan sulfate in a dose-dependent pattern.

Arshad et al. [35] presented the effect of different biotic and abiotic factors for active degradation of endosulfan supplemented in soils by *Pseudomonas aeruginosa*. The increase in pH of culture media up to 8 resulted in approximately 87% and 83% degradation of α- and β- endosulfan, respectively. The degradation rate of endosulfan was found to vary with the texture of loamy soils and incubation periods.

Weir et al. [36] isolated and identified *Arthrobacter* sp. KW from endosulfan enriched soil that utilized endosulfan and its toxic metabolite endosulfan sulfate as sole sulfur source. They observed that specific "ese" genes expressing the formation of monooxygenase enzyme relying on flavin was responsible for degradation of endosulfan and endosulfan sulfate. Genomic transformation and sequence-based tests through the Basic Local Alignment Search Tool (BLAST) technique encourage its concurrent use to solve the contaminant degradation problem through the application of transgenic bacteria.

Mycobacterium sp. strain ESD is a gram-positive rod-shaped bacteria that can utilize endosulfan as a major source of carbon and sulfur. The 16S rRNA testing signifies 98.3% similarity with *Mycobacterium* sp. strain LB501T. This bacterium efficiently degrades β-endosulfan and produces endosulfan monoaldehyde and endosulfan hydroxy ether as metabolic intermediate [29].

Goswami et al. [37] have reported *Bordetella* sp. B9 for degradation of α- and β- endosulfan in both liquid culture as well as soil microcosm. At 100 µg/mL concentration of endosulfan, *Bordetella* sp. B9 exhibited maximum growth and biodegradation capabilities. About 80% of α-endosulfan and 86% of β-endosulfan was degraded within 18 days of incubation periods with the production of endosulfan sulfate, endosulfan ether, and endosulfan lactone as an intermediate metabolite suggesting both oxidative and hydrolytic pathway of degradation.

3.2. Endosulfan Degradation by Fungi

Like bacterial strains, a number of fungal strains (Table 1) are also known to participate in endosulfan degradation. Investigations carried out on some of the important fungal strains causing degradation of endosulfan are described in the following paragraphs.

Goswami et al. [38] reported about 95% degradation of α-endosulfan and 97% degradation of β-endosulfan within 18 days of *Aspergillus sydoni* incubation at ambient temperature. Endosulfan sulfate, endosulfan ether, and endosulfan lactone were the important intermediate metabolites. Degradation and the concomitant increase in bacterial cell growth suggested the utilization of endosulfan as energy sources by the tested bacteria. However, the presence of excess sucrose in the medium inhibited the endosulfan degradation by bacteria.

Table 1. Endosulfan degradation potential of bacteria, fungi, microalgae and cyanobacteria with their major metabolites

Types of organisms	Name of organism	Substrate degradation	Major metabolites	Incubation period (days)	References
Bacteria	*Pseudomonas* sp. IITR01	α-endosulfan and endosulfan sulfate	Endosulfan diol, endosulfan ether, endosulfan lactone and endosulfan sulfate	12	[52]
	Mycobacterium sp.	β- endosulfan	Endosulfan hydroxyether and endosulfan monoaldehyde	10	[29]
	Alkaligens faecalis JBW$_4$	α- and β-endosulfan	Endosulfan lactone and endosulfan diol	21	[31]
	Achromobacter xylosoxidans strain C8B	α- and β-endosulfan	Endosulfan lactone, endosulfan ether and endosulfan sulfate	15	[27]
	Agrobacterium tumifaciens EBRI25 and *A. tumifaciens* PT-3	α- and β-endosulfan	Endosulfan sulfate	5-7	[28]
	Klebsiella oxytosa KE-8	α- and β-endosulfan	Endosulfan sulfate	7-12	[34]

Table 1. (Continued)

Types of organisms	Name of organism	Substrate degradation	Major metabolites	Incubation period (days)	References
	Bacillus megaterium KKC₇	α- and β- endosulfan	Endosulfan sulfate	-	[32]
	Pseudomonas aeruginosa	α- and β- endosulfan	-	7-8	[35]
	Arthrobacter sp strain KW	α- and β- endosulfan	Endosulfan sulfate	-	[36]
	Strenotrophomonas maltophilia	α- and β- endosulfan	Endosulfan diol	14	[55]
	Bordetella sp. B9	α- and β- endosulfan	Endosulfan ether and endosulfan lactone	15-20	[37]
	Burkholderia cepacea TJ2.B6	α- and β- endosulfan	Endosulfan diol and endosulfan hydroxyether	14	[42]
	Pandoraea sp.	α- and β- endosulfan	Endosulfan diol and endosulfan ether	18	[23]
	Paenibacillus sp. ISTP10	α- and β- endosulfan	Endosulfan diol, endosulfan lactones and endosulfan ether	10	[54]
Fungi	*Aspergillus sydoni*	α- and β- endosulfan	Endosulfan sulfate, endosulfan ether and endosulfan lactone	18	[38]

Types of organisms	Name of organism	Substrate degradation	Major metabolites	Incubation period (days)	References
	Lasiodiplodia sp. JAS12	α- and β- endosulfan	Endosulfan sulfate	12	[33]
	Mortierella sp. strain W8	α- and β- endosulfan	Endosulfan diol, endosulfan ether and endosulfan lactone	28	[39]
	Trichoderma harzianum 2023	α- and β- endosulfan	Endosulfan diol and endosulfan sulfate	10-13	[40]
	Aspergillus niger	α- and β- endosulfan	Endosulfan diol and endosulfan sulfate	12	[53]
	Aspergillus terricola, A. terreus and *Chaetosartorya stromatoides*	α- and β- endosulfan	Endosulfan hydroxyether and endosulfan sulfate	10-12	[42]
	Penicillium sp. CHE23	α- and β- endosulfan	-	2-6	[52]
	Fusarium ventricosum	α- and β- endosulfan	-	15	[23]
	Mucor thermohyalospora MTCC1384	α- and β- endosulfan	Endosulfan sulphate and endosulfan diol	20	[43]

Table 1. (Continued)

Types of organisms	Name of organism	Substrate degradation	Major metabolites	Incubation period (days)	References
	Cladosporium oxysporum Botryosphaeria laricina JAS6 and *Aspergillus tamarii JAS9*	α- and β- endosulfan	Endosulfan diol and endosulfan ether	10	[45]
	White rot and Brown rot fungi	α- and β- endosulfan	Endosulfan sulphate and endosulfan ether	1	[46]
Microalgae/Cyanobacteria	*Chlorococcum* and *scenedesmus spp.*	α- endosulfan	Endosulfan sulphate and endosulfan ether	20	[47]
	Anabaena sp. PCC 7120 and *Anabaena flos-aquae*	α- and β- endosulfan	Endosulfan endodiol	15	[48]
	Nostoc muscorum, Anabaena variabilis and *Aulosira fertilissima*	α- and β- endosulfan	-	-	[49]

Lasiodiplodia sp. JAS12 was described to actively degrade α- and β- endosulfan through oxidative metabolic processes. The biodegradation process followed the first order reaction kinetics. The degradation was confirmed by spectroscopic investigations. Fourier transform infrared spectroscopy (FTIR) exhibited specific peaks at 1400 and 1402 cm^{-1} corresponding to –COOH group suggesting active degradation pattern [33]. The potential abilities of tested fungi, however, could be improved through the development of transgenics possessing the genes participating in endosulfan degradation for management of soil and water contaminated with different isomers of endosulfan.

Kataoka and his co-workers [39] isolated *Mortierella* sp. strain W8 from buried sites enriched with persistent organic pollutants and studied endosulfan degradation potential through soil microcosm assay. Gas chromatography-Mass spectroscopy (GC-MS) analyses of extracted samples detected endosulfan diol, endosulfan ether, endosulfan lactone, and endosulfan sulfate as an intermediate degradation product. Approximately 70% of α- endosulfan and 50% of β- endosulfan was degraded after 21-28 days incubation periods.

Katayama and Matsumura [40] have reported the involvement of sulfatase and phosphatase activity of *Trichoderma harzianum* 2023 for active degradation of α-endosulfan and β-endosulfan. The endosulfan had no apparent effect on growth and biomass synthesis of the fungi selected for endosulfan degradation suggesting their utilization for degradation of contaminant even at higher concentration. The metabolic degradation of endosulfan corresponded to oxidative degradation mechanism.

Bhalerao [41] studied the tolerance and biodegradation of endosulfan in enrichment culture. Degradation metabolites were identified from acetonitrile fraction through HPLC, GC-MS and FTIR analysis. The major intermediary metabolites produced after 12 days of incubation with *Aspergillus niger* ARIFCC 1053 were identified as endosulfan sulfate and endosulfan diol.

Hussain et al. [42] described that fungal consortium of *Aspergillus terricola, A. terreus* and *Chaetosartorya stromatoides* can degrade α- as well as β- endosulfan up to 75% and 20%, respectively decreasing medium pH to

7.0 and 3.2, respectively. The optimum biodegradation was achieved at ambient temperature with slightly agitating conditions. Higher incubation temperature improved the degradation of endosulfan at varying cellular densities of the above-mentioned consortium of fungal species.

Fusarium ventricosum utilizes endosulfan as the sulfur source and poor source of electrons for reducing energy. In a study conducted by Siddique et al. [23], it was reported that culture medium became acidic due to endosulfan dehalogenation and production of different organic acids after catabolism. Rapid degradation of endosulfan leading to the formation of some common metabolic by-products (endosulfan diol, endosulfan ether, and endosulfan lactone) may help in commercial application of *Fusarium ventricosum* for bioremediation of endosulfan polluted areas. They opined that even 100% degradation of α- and β- endosulfan (100 mg/L) could be made possible after 12 days of incubation.

Shetty et al. [43] reported biotransformation and biodegradation of endosulfan and identified biodegradation intermediates as endosulfan sulfate and endosulfan diol. They concluded that approximately 70% of α- and 50% of β- endosulfan can be degraded within 20 days of incubation periods.

The fungus *Cladosporium oxysporum* was isolated from the egg of *Meloidogyne incognita*, a root-knot nematode [44] and evaluation was performed to check its efficiency for endosulfan degradation. They have described significant degradation (~89%) of endosulfan as compared to un-inoculated control (~66.8%) within 15 days of incubation period.

Silambarasan and Abraham [45] have reported the isolation of *Botryosphaeria laricina* JAS6 and *Aspergillus tamarii* JAS9 from *Abelmoschus esculentus* and their endosulfan degradation potential for the first time. The isolated strains were able to survive at endosulfan concentration equivalent to 1300 mg/L and exhibited excellent growth at 1000 mg/L. The degradation of endosulfan by tested fungi lowered the media pH from 6.8 to 3.4 and 3.8 as noted for *Botryosphaeria laricina* JAS6 and *Aspergillus tamarii* JAS9, respectively.

Trametes hirsuta, *T. versicolor*, and *Pleurotus ostreatus*, commonly called as "white rot fungi" may actively participate in the degradation of varieties of organic pesticides through their metabolic enzymes. The different oxidative exoenzymes of lignin degradation are known to be involved in complete mineralization of endosulfan followed by formation of toxic intermediate endosulfan sulfate and endosulfan ether. More specifically, the basidiomycetous peroxidases and laccase are important enzymatic tools for the degradation of several organic pollutants. About 4-10% of organic pollutants (endosulfan) can be removed within 4-24h after incubation with white and brown rot fungi [46].

3.3. Endosulfan Degradation by Microalgae and Cyanobacteria

Different types of phototropic microalgae and cyanobacteria have been commonly utilized for the treatment of sewage sludge due to their degradation potentiality of various organic dyes, pesticides and other persistent organic pollutants. The *Chlorococcum, Scenedesmus, Anabaena, Nostoc,* and *Aulosira* are some of the efficient algae which have been recently used for the degradation of endosulfan (Table 1).

Application of different algal species could help in remediation of endosulfan contaminated soil and water. Species of *Chlorococcum* and *Scenedesmus* have been described to degrade 65-70% α-endosulfan within 20 days, while 90-95% degradation was evidenced after 30 days in liquid medium [47]. The tested algae produced endosulfan sulfate as the major metabolite while endosulfan ether as minor metabolite without any significant amount was also documented. A very interesting feature of the investigation of algal degradation was presumably efficient degradation of endosulfan sulfate (30 μg/L) at a higher rate as compared to α-endosulfan. Light-dependent hydrolysis of endosulfan at neutral or alkaline pH through algal metabolic activities effectively degrades and detoxifies endosulfan from soil and water. Some of the important algal species tested for their endosulfan degradation abilities by different researchers are being described here.

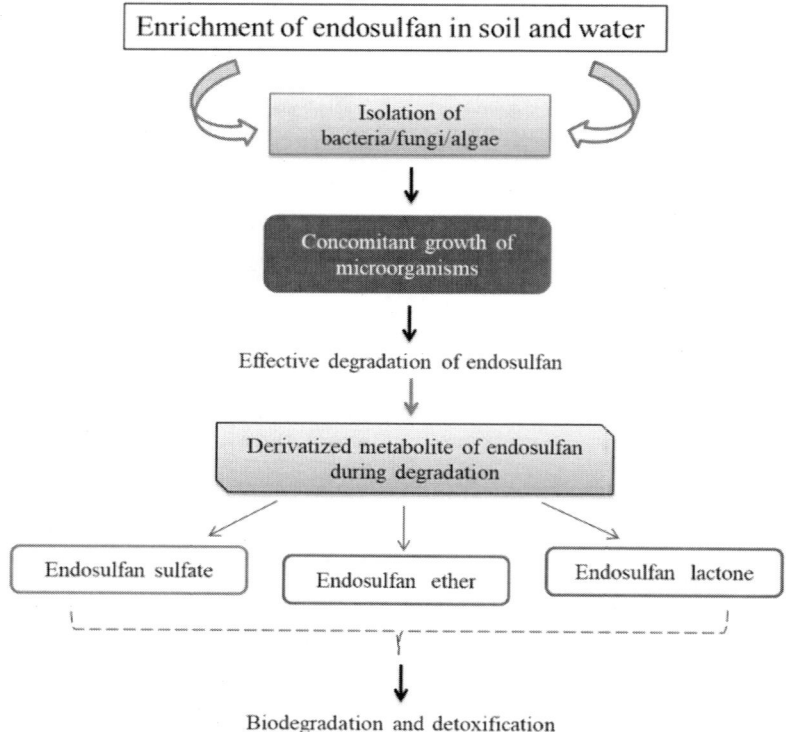

Figure 2. The schematic pathways of endosulfan degradation and by-product formation by bacteria, fungi and microalgae.

Two different species of *Anabaena* utilized varieties of metabolic activities for the degradation of endosulfan. Basically, they were involved in endosulfan detoxification through the formation of non-toxic endodiol. Concomitant growth of bacterial species together with the formation of endosulfan monoaldehyde suggests a novel method of oxidative and hydrolytic degradation facilitated by specific enzymes depending on the availability of oxygen [48].

Kumar et al. [8] reported varying degradation potential and tolerance mechanism of *Nostoc muscorum*, *Anabaena variabilis* and *Aulosira fertilissima* when exposed to 10 μg/mL concentration of endosulfan. At high concentrations of endosulfan, the growth of the cyanobacteria was reduced thus representing toxic effects of endosulfan on vital cellular processes.

Furthermore, endosulfan toxicity effectively alters the pigment profile, carbohydrate and protein metabolism of *Nostoc muscorum*, *Anabaena variabilis* and *Aulosira fertilissima*. However, an increase in intracellular proline content can significantly help to cope up the toxicity and facilitate possible oxidative detoxification process. The schematic representation of endosulfan degradation and by-product formation by bacteria, fungi, and microalgae is presented in Figure 2.

4. Genetics and Molecular Mechanism for Endosulfan Detoxification

A number of studies have reported the detoxification of endosulfan by utilization of plethora of microbes with the generation of toxic endosulfan sulfate (oxidation process) or comparatively less toxic endosulfan diol through hydrolytic process [49, 50]. In recent years, environmental biologists are especially focused on the specific genes of microbes having the potential of endosulfan degradation through its protein or enzymatic products. Recent studies have indicated either direct involvement or alteration of endosulfan metabolizing genes (ESD genes) through significant changes in the expression of mRNA or protein products [30]. Bajaj et al. [51] attempted to identify the role of the monooxygenase genes for endosulfan degradation in *Pseudomonas* sp. strain IITR01. An interesting finding from the *Pseudomonas* sp. was the absence of monooxygenase "esd" gene suggesting the direct transformation of endosulfan into endosulfan lactone without ether formation. SDS-PAGE and native gel analysis revealed a well-characterized protein of 150 KDa for α-endosulfan degradation. Proteins from different fungal species engaged in endosulfan degradation are reported at a lesser extent but the horizontal transfer of such catabolic genes into the bacterial system is not explored till date. Sutherland and his co-workers [29] have reported the involvement of monooxygenase and flavin reductase genes for degradation of β-endosulfan by *Mycobacterium* sp ESD. Weir et al. [36] observed active degradation of α and β endosulfan to endosulfan monoalcohol and endosulfan sulfate to produce endosulfan hemisulphate through "use" monooxygenase gene. Different recombinant strains of *Mycobacterium smegmatis* such as flavin reductase (FR), flavin mononucleotide (FMN), flavin adenine dinucleotide (FAD) and riboflavin are known to be utilized as an electron donor for endosulfan degradation. Sulfur regulation for endosulfan detoxification occurred in *M. smegmatis* through Apa I fragment (3 kb) with "esd" constitutive promoter and 53 bp palindromic sequence originated upstream of the responsible gene. However, the involvement of specific genes and their protein or enzymatic products for endosulfan as well as its inter-metabolic degradation products require much research inputs in this direction. An illustration of different genes participating in endosulfan degradation together with the detection of metabolic products using modern analytical techniques is represented in Figure 3.

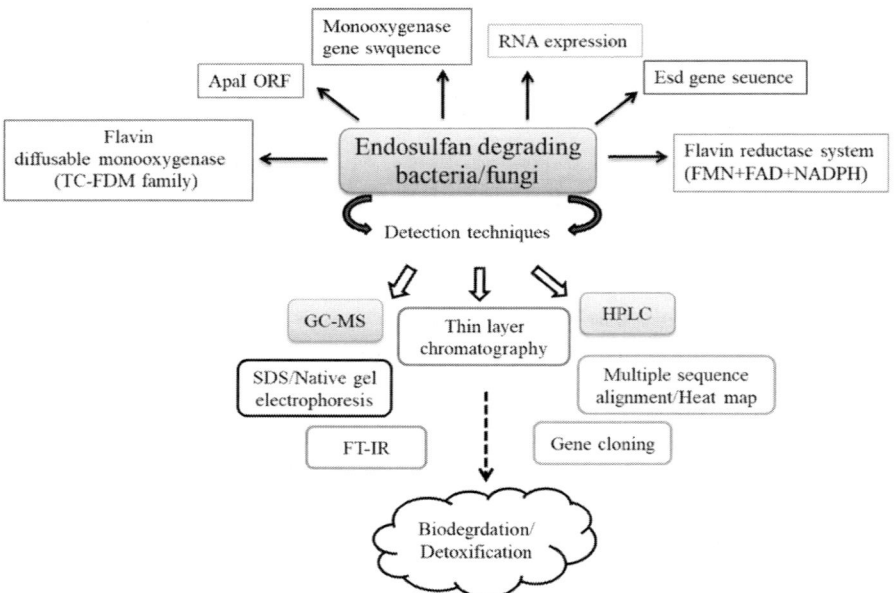

Figure 3. Different genes participating in endosulfan degradation together with the detection of metabolic products using modern analytical techniques.

CONCLUSION AND FUTURE PERSPECTIVES

The huge application of endosulfan in agricultural fields for the management of crop productivity has contaminated the soil and water at varying degrees. The toxicity of endosulfan is evidently reported to human health and the natural environment. The application of diverse microbes for degradation of endosulfan to non-toxic products is an eco-friendly and less expensive approach for remediation of contaminated sites. So far, a number of bacteria, fungi, and algae isolated and identified from endosulfan contaminated sites has been documented. Although, laboratory studies have shown promising results for their degradation, however, field application showing good results are scanty. Identification of different genes and proteins from different organisms taking part in the degradation of endosulfan would give a new direction for efficient degradation of endosulfan. Further, transgenic microbes harboring the genes involved in endosulfan degradation could also be considered for rapid degradation of endosulfan under changing environmental conditions. However, prior to their field application, safety issues must be taken into account.

REFERENCES

[1] Çok, İ., Yelken, Ç., Durmaz, E., Üner, M., Sever, B. and Satır, F., 2011. "Polychlorinated biphenyl and organochlorine pesticide levels in human breast milk from the Mediterranean city Antalya, Turkey." *Bulletin of environmental contamination and toxicology*, 86(4), 423-427.

[2] Maier-Bode, H., 1968. "Properties, effect, residues and analytics of the insecticide endosulfan". In *Residue Reviews/Rückstands-Berichte* (pp. 1-44). Springer, New York, NY.

[3] Agnihotri, A. G., 1999. "Pesticide: safety evaluation and monitoring." Indian Agricultural Research Institute, Division of Agricultural Chemicals.

[4] Chan, M. P., Morisawa, S., Nakayama, A., Kawamoto, Y. and Yoneda, M., 2006. "Development of an in vitro blood–brain barrier model to study the effects of endosulfan on the permeability of tight junctions and a comparative study of the cytotoxic effects of endosulfan on rat and human glial and neuronal cell cultures". *Environmental Toxicology: An International Journal*, 21(3), 223-235.

[5] Fernandez, M. F., Olmos, B., Granada, A., López-Espinosa, M. J., Molina-Molina, J. M., Fernandez, J. M., Cruz, M., Olea-Serrano, F. and Olea, N., 2007. "Human exposure to endocrine-disrupting chemicals and prenatal risk factors for cryptorchidism and hypospadias: a nested case-control study." *Environmental health perspectives*, 115(Suppl 1), 8.

[6] Bajpayee, M., Pandey, A. K., Zaidi, S., Musarrat, J., Parmar, D., Mathur, N., Seth, P. K. and Dhawan, A., 2006. "DNA damage and mutagenicity induced by endosulfan and its metabolites." *Environmental and Molecular Mutagenesis*, 47(9), 682-692.

[7] US EPA 2002. "Reregistration eligibility decision for endosulfan". Report No. 738-R-02-013. United States Environmental Protection Agency, Washington, D.C., USA. Available at: www.epa.gov/oppsrrd/REDs/endosulfan_red.pdf (accessed Dec 1, 2012)

[8] Kumar, S., Habib, K. and Fatma, T., 2008. "Endosulfan induced biochemical changes in nitrogen-fixing cyanobacteria." *Science of the Total Environment*, 403(1-3), 130-138.

[9] Stanley, K. A., Curtis, L. R., Simonich, S. L. M. and Tanguay, R. L., 2009. "Endosulfan I and endosulfan sulfate disrupts zebrafish embryonic development." *Aquatic toxicology*, 95(4), 355-361.

[10] Atsdr, U. S., 1997. "*Agency for Toxic Substances and Disease Registry*" Case Studies in environmental medicine. http://www. atsdr. cdc. gov/HEC/CSEM/csem. html.

[11] Menezes, R. G., Qadir, T. F., Moin, A., Fatima, H., Hussain, S. A., Madadin, M., Pasha, S. B., Al Rubaish, F. A. and Senthilkumaran, S., 2017. "Endosulfan poisoning: an overview." *Journal of forensic and legal medicine*, 51, 27-33.

[12] Enhui, Z., Na, C., MengYun, L., Jia, L., Dan, L., Yongsheng, Y., Ying, Z. and DeFu, H., 2016. "Isomers and their metabolites of endosulfan induced cytotoxicity and oxidative damage in SH-SY 5 Y cells." *Environmental toxicology*, 31(4), 496-504.

[13] Lu, Y., Morimoto, K., Takeshita, T., Takeuchi, T. and Saito, T., 2000. "Genotoxic effects of alpha-endosulfan and beta-endosulfan on human HepG2 cells". *Environmental Health Perspectives*, 108(6), 559-561.

[14] Jayachandra, S., 2016. "Impact of Endosulfan on Male Reproductive functions-a review." *Perspecive in Medical Research*. 4, 44-49.

[15] Abalis, I. M., Eldefrawi, M. E. and Eldefrawi, A. T., 1986. "Effects of insecticides on GABA-induced chloride influx into rat brain microsacs." *Journal of Toxicology and Environmental Health, Part A Current Issues*, 18(1), 13-23.

[16] Du, H., Wang, M., Dai, H., Hong, W., Wang, M., Wang, J., Weng, N., Nie, Y. and Xu, A., 2015. "Endosulfan isomers and sulfate metabolite induced reproductive toxicity in *Caenorhabditis elegans* involves genotoxic response genes." *Environmental science & technology*, 49(4), 2460-2468.

[17] Saiyed, H., Dewan, A., Bhatnagar, V., Shenoy, U., Shenoy, R., Rajmohan, H., Patel, K., Kashyap, R., Kulkarni, P., Rajan, B. and Lakkad, B., 2003. "Effect of Endosulfan on male reproductive development." *Environmental Health Perspectives*, 111(16), 1958-1962.

[18] Knoevenagel, K. and Himmelreich, R., 1976. Degradation of compounds containing carbon atoms by photooxidation in the presence of water. *Archives of environmental contamination and toxicology*, 4(1), 324-333.

[19] Kwon, G. S., Sohn, H. Y., Shin, K. S., Kim, E. and Seo, B. I., 2005. "Biodegradation of the organochlorine insecticide, endosulfan, and the toxic metabolite, endosulfan sulfate, by *Klebsiella oxytoca* KE-8." *Applied Microbiology and Biotechnology*, 67(6), 845-850.

[20] Sutherland, T. D., Horne, I., Lacey, M. J., Harcourt, R. L., Russell, R. J. and Oakeshott, J. G., 2000. "Enrichment of an endosulfan-degrading

mixed bacterial culture." *Applied and Environmental Microbiology*, 66(7), 2822-2828.

[21] Jesitha, K. and Harikumar, P. S., 2019. Development of a bioreactor system for the remediation of endosulfan. *H2Open Journal*, 1-9.

[22] Singh, N. S., Sharma, R. and Singh, D. K., 2019. "Identification of enzyme (s) capable of degrading endosulfan and endosulfan sulfate using in silico techniques." *Enzyme and Microbial Technology*, 124, 32-40.

[23] Siddique, T., Okeke, B. C., Arshad, M. and Frankenberger, W. T., 2003. "Enrichment and isolation of endosulfan-degrading microorganisms." *Journal of Environmental Quality*, 32(1), 47-54.

[24] Sachdev, D. P. and Cameotra, S. S., 2013. "Biosurfactants in agriculture". *Applied Microbiology and Biotechnology*, 97(3), 1005-1016.

[25] Jayashree, R. and Vasudevan, N., 2009. "Effect of Tween 80 and moisture regimes on endosulfan degradation by *Pseudomonas aeruginosa*." *Applied Ecology and Environmental Research*, 7(1), 35-44.

[26] Kwon, G. S., Kim, J. E., Kim, T. K., Sohn, H. Y., Koh, S. C., Shin, K. S. and Kim, D. G., 2002. "*Klebsiella pneumoniae* KE-1 degrades endosulfan without formation of the toxic metabolite, endosulfan sulfate." *FEMS Microbiology Letters*, 215(2), 255-259.

[27] Singh, N. S. and Singh, D. K., 2011. "Biodegradation of endosulfan and endosulfan sulfate by *Achromobacter xylosoxidans* strain C8B in broth medium." *Biodegradation*, 22(5), 845-857.

[28] Thangadurai, P. and Suresh, S., 2014. Biodegradation of endosulfan by soil bacterial cultures. *International Biodeterioration & Biodegradation*, 94, 38-47.

[29] Sutherland, T. D., Horne, I., Harcourt, R. L., Russell, R. J. and Oakeshott, J. G., 2002. "Isolation and characterization of a *Mycobacterium* strain that metabolizes the insecticide endosulfan." *Journal of applied microbiology*, 93(3), 380-389.

[30] Verma, A., Ali, D., Farooq, M., Pant, A. B., Ray, R. S. and Hans, R. K., 2011. "Expression and inducibility of endosulfan metabolizing

gene in *Rhodococcus* strain isolated from earthworm gut microflora for its application in bioremediation." *Bioresource Technology*, 102(3), 2979-2984.

[31] Kong, L., Zhu, S., Zhu, L., Xie, H., Su, K., Yan, T., Wang, J., Wang, J., Wang, F. and Sun, F., 2013. "Biodegradation of organochlorine pesticide endosulfan by bacterial strain *Alcaligenes faecalis* JBW4." *Journal of Environmental Sciences*, 25(11), 2257-2264.

[32] Seralathan, M. V., Sivanesan, S., Bafana, A., Kashyap, S. M., Patrizio, A., Krishnamurthi, K. and Chakrabarti, T., 2014. "Cytochrome P450 BM3 of *Bacillus megaterium*—A possible endosulfan biotransforming gene." *Journal of Environmental Sciences*, 26(11), 2307-2314.

[33] Abraham, J. and Silambarasan, S., 2014. "Role of novel fungus *Lasiodiplodia* sp. JAS12 and plant growth promoting bacteria *Klebsiella pneumoniae* JAS8 in mineralization of endosulfan and its metabolites." *Ecological Engineering*, 70, 235-240.

[34] Kwon, G. S., Sohn, H. Y., Shin, K. S., Kim, E. and Seo, B. I., 2005. "Biodegradation of the organochlorine insecticide, endosulfan, and the toxic metabolite, endosulfan sulfate, by *Klebsiella oxytoca* KE-8." *Applied Microbiology and Biotechnology*, 67(6), 845-850.

[35] Arshad, M., Hussain, S., and Saleem, M., 2008. Optimization of environmental parameters for biodegradation of alpha and beta-endosulfan in soil slurry by *Pseudomonas aeruginosa*. *Journal of Applied Microbiology*, 104(2), 364-370.

[36] Weir, K. M., Sutherland, T. D., Horne, I., Russell, R. J. and Oakeshott, J. G., 2006. "A single monooxygenase, ese, is involved in the metabolism of the organochlorides endosulfan and endosulfan in an *Arthrobacter* sp." *Applied and Environmental Microbiology*, 72(5), 3524-3530.

[37] Goswami, S. and Singh, D. K., 2009. "Biodegradation of α and β endosulfan in broth medium and soil microcosm by bacterial strain *Bordetella* sp. B9." *Biodegradation*, 20(2), 199-207.

[38] Goswami, S., Vig, K., and Singh D.K., 2009. "Biodegradation of α and β endosulfan by *Aspergillus sydoni*." *Chemosphere* 75(7), 883-888.

[39] Kataoka, R., Takagi, K. and Sakakibara, F., 2010. "A new endosulfan-degrading fungus, *Mortierella* species, isolated from a soil contaminated with organochlorine pesticides." *Journal of Pesticide Science*, 35(3), 326-332.

[40] Katayama, A. and Matsumura, F., 1993. "Degradation of organochlorine pesticides, particularly endosulfan, by *Trichoderma harzianum*." *Environmental Toxicology and Chemistry: An International Journal*, 12(6), 1059-1065.

[41] Bhalerao, T. S. (2013). "Biominerlization and possible endosulfan degradation pathway adapted by *Aspergillus niger*." *Journal of Microbiology and Biotechnology*, 23(11), 1610-1616.

[42] Hussain, S., Arshad, M., Saleem, M. and Zahir, Z. A., 2007. "Screening of soil fungi for in vitro degradation of endosulfan." *World Journal of Microbiology and Biotechnology*, 23(7), 939-945.

[43] Shetty, P. K., Mitra, J., Murthy, N. B. K., Namitha, K. K., Savitha, K. N. and Raghu, K., 2000. "Biodegradation of cyclodiene insecticide endosulfan by *Mucor thermohyalospora* MTCC 1384." *Current Science*, 1381-1383.

[44] Mukherjee, I. and Mittal, A., 2005. "Bioremediation of endosulfan using *Aspergillus terreus* and *Cladosporium oxysporum*." *Bulletin of Environmental Contamination and Toxicology*, 75(5), 1034-1040.

[45] Silambarasan, S. and Abraham, J., 2013. "Mycoremediation of endosulfan and its metabolites in aqueous medium and soil by *Botryosphaeria laricina* JAS6 and *Aspergillus tamarii* JAS9." *PloS One*, 8(10), e77170.

[46] Ulčnik, A., Cigić, I. K. and Pohleven, F., 2013. Degradation of lindane and endosulfan by fungi, fungal and bacterial laccases. *World Journal of Microbiology and Biotechnology*, 29(12), 2239-2247.

[47] Sethunathan, N., Megharaj, M., Chen, Z. L., Williams, B. D., Lewis, G. and Naidu, R., 2004. "Algal degradation of a known endocrine-disrupting insecticide, α-endosulfan, and its metabolite, endosulfan sulfate, in liquid medium and soil." *Journal of Agricultural and Food chemistry*, 52(10), 3030-3035.

[48] Lee, S. E., Kim, J. S., Kennedy, I. R., Park, J. W., Kwon, G. S., Koh, S. C. and Kim, J. E., 2003. Biotransformation of an organochlorine insecticide, endosulfan, by *Anabaena* species. *Journal of Agricultural and Food chemistry*, 51(5), 1336-1340.

[49] Bumpus, J. A. and Aust, S. D., 1987. "Biodegradation of environmental pollutants by the white rot fungus *Phanerochaete chrysosporium*: involvement of the lignin-degrading system." *BioEssays*, 6(4), 166-170.

[50] Kullman, S. W. and Matsumura, F., 1996. "Metabolic pathways utilized by *Phanerochaete chrysosporium* for degradation of the cyclodiene pesticide endosulfan." *Applied and Environmental Microbiology*, 62(2), 593-600.

[51] Bajaj, A., Pathak, A., Mudiam, M. R., Mayilraj, S. and Manickam, N., 2010. "Isolation and characterization of a *Pseudomonas* sp. strain IITR01 capable of degrading α-endosulfan and endosulfan sulfate." *Journal of Applied Microbiology*, 109(6), 2135-2143.

[52] Romero-Aguilar, M., Tovar-Sánchez, E., Sánchez-Salinas, E., Mussali-Galante, P., Sánchez-Meza, J. C., Castrejón-Godínez, M. L., Dantán-González, E., Trujillo-Vera, M. Á. and Ortiz-Hernández, M. L., 2014. "*Penicillium* sp. as an organism that degrades endosulfan and reduces its genotoxic effects." *SpringerPlus*, 3(1), 536-546.

[53] Bhalerao, T. S. and Puranik, P. R., 2007. "Biodegradation of organochlorine pesticide, endosulfan, by a fungal soil isolate, *Aspergillus niger*." *International Biodeterioration & Biodegradation*, 59(4), 315-321.

[54] Kumari, M., Ghosh, P. and Thakur, I. S., 2014. "Microcosmic study of endosulfan degradation by *Paenibacillus* sp. ISTP10 and its toxicological evaluation using mammalian cell line." *International Biodeterioration & Biodegradation*, 96, 33-40.

[55] Kumar, K., Devi, S. S., Krishnamurthi, K., Kanade, G. S. and Chakrabarti, T., 2007. "Enrichment and isolation of endosulfan degrading and detoxifying bacteria." *Chemosphere*, 68(2), 317-322.

ABOUT THE EDITOR

Ishwar Chandra Yadav, PhD
Research Scientist, Tokyo University of Agriculture and Technology,
Tokyo, Japan

Ishwar Chandra Yadav received his Ph.D. in Environmental Science from Banaras Hindu University, Varanasi, India in 2012. Currently, He is working as Research Scientist at Tokyo University of Agriculture Science and Technology, Tokyo, Japan. He has published a number of articles in high impact factor scientific journals. His research interests are biogeochemistry of persistent organic pollutant, brominated and organophosphate flame retardants, carbonaceous aerosols and PM 2.5. He is associate editors of Science of the Total Environment and Data in Brief.

Ningombam Linthoingambi Devi, PhD
Assistant Professor, Central University of South Bihar,
Gaya, India

Ningombam Linthoingambi Devi obtained her Ph.D. in Environmental Engineering from China University of Geosciences, Wuhan China in 2011. At present, she is an assistant professor in the School of Earth, Biological and Environmental Sciences, Central University of South Bihar, Gaya,

India. She has published a number of articles in international peer-reviewed journals. Her main research interests are biogeochemistry of aerosols, PM2.5, black carbon, heavy metal pollution, and persistent organic pollutants

INDEX

#

2nd highest among the South Asian, 180

A

abiotic parameter, 226
access, 70, 75, 212
accidental, 60, 84, 152, 156, 210
acetic acid, 193
acetone, 85, 88, 90, 91, 92, 93, 94, 97, 100, 101, 102, 103, 104, 105, 107
acetonitrile, 92, 93, 94, 95, 105, 259
acetylcholinesterase, 32, 45, 46
acid, 6, 10, 45, 83, 86, 88, 156, 160, 173, 193, 202, 217, 236, 249
acidic, 143, 211, 218, 226, 260
activated carbon, 218
active oxygen, 237
adipose tissue, 82, 85, 86, 98, 100, 108, 110, 112, 120, 126, 133, 135, 137, 138
Adopt, 194
adsorption, 54, 88, 108, 109, 148, 191, 218, 219
adverse effects, 1, 25, 33, 157, 188

agriculture, 2, 7, 26, 34, 38, 48, 49, 56, 70, 72, 78, 79, 81, 150, 175, 177, 179, 182, 186, 189, 192, 194, 195, 200, 209, 268
agriculture sector in pakistan, 175
Agrobacterium, 252, 255
air, 3, 4, 9, 11, 12, 13, 26, 27, 28, 29, 34, 35, 38, 39, 40, 41, 43, 49, 53, 54, 58, 60, 76, 77, 82, 143, 147, 149, 161, 174, 180, 205, 214, 220, 226, 228, 231, 232, 233, 243, 248
air temperature, 149
algae, 12, 151, 222, 248, 261, 265
alkaline phosphatase, 31
alpines, 161
amino, 32, 193, 224, 249
amino acid, 32, 193, 224
amphibians, 32, 238
animal disease, 83
apoptosis, 164, 214, 235
aquatic life, 143, 152, 153, 156, 162, 183
aquatic systems, 34
aqueous solutions, 191, 202, 243
arrest, 109
Asia, 2, 10, 31, 114, 164, 195
Aspergillus terreus, 223, 244, 249, 270

assessment, 43, 48, 51, 77, 78, 126, 138, 139, 174, 194, 198, 199, 213, 219, 220, 229, 233
atmosphere, 2, 13, 19, 26, 28, 29, 38, 39, 41, 42, 47, 49, 58, 76, 142, 148, 150, 161, 167, 212, 216, 231
atmospheric deposition, 150, 152, 210

B

bacteria, 82, 145, 165, 208, 216, 221, 222, 224, 232, 240, 241, 243, 248, 249, 251, 254, 255, 262, 263, 265, 269, 271
bacterial cells, 149
bacterial colonies, 12
bacterial strains, 191, 244, 254
bacterium, 67, 165, 192, 222, 252, 254
ban, 55, 56, 57, 59, 63, 64, 72, 73, 150, 178, 179, 182, 206, 209, 213
Bangladesh, 3, 14, 57, 178
basophils, 158
benefits, 70, 195
bioaccumulation, 2, 6, 15, 18, 25, 26, 27, 81, 143, 155, 164, 182, 206, 208, 212, 229
bioassay, 159, 186, 190
biochemical processes, 216
biodegradation, viii, 149, 165, 167, 192, 193, 203, 206, 208, 209, 221, 222, 223, 230, 231, 232, 240, 241, 243, 244, 247, 248, 251, 252, 253, 254, 259, 260, 267, 268, 269, 270, 271
biological activities, 151, 216
biological processes, 206
biological samples, 82, 84, 85, 109, 110, 129, 130, 132, 135, 137, 183
biomass, 70, 160, 173, 225, 253, 259
biomolecules, 213
biomonitoring, 82, 84, 111, 113, 126, 138, 139

bioremediation, 202, 241, 243, 251, 253, 260, 269
biotic, 29, 38, 145, 161, 221, 253
birds, 12, 33, 41
birth weight, 134
bleeding, 131, 134
blindness, 158
blood, 14, 31, 32, 63, 82, 84, 85, 87, 92, 93, 94, 97, 108, 109, 110, 111, 116, 117, 118, 119, 126, 127, 129, 130, 133, 134, 135, 136, 137, 139, 140, 143, 158, 163, 172, 186, 198, 199, 200, 214, 248, 266
blood plasma, 14
blood pressure, 31
bone, 214, 236
bone marrow, 214, 236
brain, 2, 31, 32, 33, 45, 47, 158, 266, 267
brain damage, 31
Brazil, 2, 8, 16, 105, 111, 115, 119, 124, 134, 135, 138, 145, 147, 148, 167
breakdown, 13
breast cancer, 10
breast milk, 82, 85, 86, 103, 110, 114, 115, 123, 126, 130, 131, 132, 138, 265
breastfeeding, 114
Brittany, 96
Burkina Faso, 160, 174
by-products, 218, 220, 249, 260

C

cabbage, 71
cadmium, 21, 128
calibration, 170, 211
Cambodia, 14, 178
cancer, 11, 31, 63, 112, 118, 129, 136, 176, 179, 214
cancers, 10
capacity building, 73, 74

carbon, 27, 29, 60, 108, 149, 202, 218, 219, 220, 221, 222, 224, 225, 226, 239, 251, 252, 254, 267
carcinogenicity, 22
cardiovascular disorders, 213
Caribbean, 9, 15, 21, 24, 155, 163
case studies, 183
case study, 41, 55, 189, 201
catalyst, 220
catchments, 151
cell culture, 266
cell division, 214, 215
cell line, 143, 214, 271
cellular communications, 213
central nervous system, 10, 31, 84, 156, 217, 248, 249
central nervous system (CNS), 31, 248, 249
cerebral palsy, 10, 157
cervix, 112, 118, 136
changing environment, 265
chemical, 4, 5, 6, 15, 26, 27, 33, 45, 64, 65, 68, 73, 74, 81, 130, 143, 145, 151, 157, 164, 176, 178, 188, 194, 206, 208, 217, 218, 249
chemical characteristics, 6
chemical degradation, 145
chemical properties, 5, 33, 218
chemicals, 15, 17, 19, 20, 42, 47, 70, 72, 79, 83, 113, 115, 145, 151, 177, 180, 182, 189, 206
children, 63, 72, 111, 113, 134, 157, 183, 214, 236, 250
China, 2, 7, 9, 18, 19, 76, 91, 96, 102, 106, 111, 113, 114, 118, 122, 124, 132, 134, 137, 139, 145, 161, 182, 210, 213, 227, 232, 234, 249
chlorinated hydrocarbons, 252
chlorine, 16, 177, 179, 186, 206, 220
chloroform, 6, 85, 92, 99, 104, 159, 179
chromatography, 82, 85, 102, 104, 109, 128, 129, 130, 132, 133, 134, 140, 188, 210, 231, 233, 239, 252, 259

chromosomal abnormalities, 2
chronic kidney disease, 117, 135, 136
cities, 34, 47, 49, 58, 60, 76, 77, 209, 231
classes, 138, 188, 237
cleanup, 84, 85, 88, 89, 90, 91, 92, 93, 94, 95, 96, 97, 98, 99, 100, 101, 102, 103, 104, 105, 106, 107, 108, 109, 137, 242
cloning, 165
clothing, 74
coastal ecosystems, 169
coastal region, 93, 130
coffee, 147, 177, 210, 216, 248
Colombia, 178
combined effect, 148, 202
commercial, 56, 71, 109, 110, 142, 155, 191, 260
communities, 21, 75
community, 54, 69, 70, 74, 79, 119, 127, 131, 135, 180, 181, 195
compliance, 75, 180
composition, 168, 236, 242
compounds, 26, 27, 40, 53, 81, 83, 84, 85, 109, 110, 112, 113, 115, 126, 130, 147, 149, 178, 218, 234, 241, 252, 253, 267
conditioning, 108
conference, 233
congenital malformations, 63
conspiracy, 19, 56
Constitution, 71
consumers, 34, 180, 248
consumers of pesticides, 180
consumption, 7, 8, 13, 32, 82, 155, 180, 182, 195
contact time, 191, 193
contaminant, 4, 12, 32, 143, 151, 207, 212, 239, 248, 254, 259
contaminated food, 9, 82
contaminated sites, 127, 206, 214, 219, 247, 265
contaminated soil, 29, 192, 193, 209, 211, 237, 239, 242, 243, 244, 249, 253, 261
contaminated soils, 193, 242, 253

contamination, ix, 11, 13, 27, 34, 38, 40, 41, 48, 50, 60, 77, 79, 114, 150, 151, 153, 168, 182, 184, 186, 187, 188, 189, 193, 194, 205, 208, 209, 210, 213, 217, 227, 228, 229, 231, 234, 248
continuous use, 162, 216
control group, 118
convention, 14, 25, 38, 73
Convention on Long-range Transboundary Air Pollution (LRTAP), 2, 14, 42, 177
correlation, 113
cost, 69, 142, 145, 191, 202, 206, 251
cotton, 2, 42, 54, 55, 60, 61, 66, 68, 71, 142, 148, 154, 157, 177, 179, 180, 181, 182, 186, 192, 196, 197, 199, 200, 201, 202, 203, 210, 216, 229, 238, 248
cotton production and pesticide use, 180
Council of Ministers, 169
Council of the European Union, 23
creatinine, 158
crop, 3, 8, 13, 70, 153, 176, 179, 180, 182, 189, 197, 198, 201, 208, 209, 214, 226, 265
crop production, 176, 179, 180, 226
crops, 2, 12, 26, 29, 46, 54, 55, 63, 69, 70, 71, 83, 127, 131, 142, 147, 148, 161, 177, 180, 186, 206, 215, 248
cryptorchidism, 113, 115, 131, 138, 266
cultivation, 60, 61, 70, 229, 248
culture, 70, 151, 163, 166, 168, 221, 230, 231, 240, 251, 253, 254, 259, 260, 268
culture media, 253
culture medium, 260
cycling, 160, 199, 214
cytotoxicity, 191, 267

D

damage, 20, 31, 45, 135, 143, 157, 163, 171, 183, 184, 199, 203, 236, 249, 266, 267
deaths, 17, 31, 39, 157, 227
decontamination, 201, 240
defects, 2, 11, 31, 113, 118, 136, 138, 139
degradation, 13, 32, 81, 126, 145, 146, 147, 148, 151, 160, 161, 166, 167, 185, 191, 192, 196, 206, 208, 209, 212, 218, 219, 220, 221, 222, 223, 224, 225, 226, 229, 230, 232, 239, 240, 241, 242, 243, 248, 251, 252, 253, 254, 255, 256, 257, 258, 259, 260, 261, 262, 263, 264, 265, 268, 270, 271
degradation process, 145, 221, 225, 226, 251
degradation rate, 206, 226, 248, 253
demonstrations, 74
Department of Commerce, 19
deposition, 4, 28, 29, 39, 41, 42, 142, 148, 158, 199
derivatives, 137, 145, 248
desorption, 202, 212
detection, 29, 82, 110, 115, 126, 129, 206, 213, 227, 237, 264
detoxification, 158, 165, 166, 221, 230, 240, 249, 251, 252, 262, 263
developed countries, 72, 177, 180
developed nations, 207
developing countries, 4, 31, 64, 70, 177, 180, 216, 236
developing nations, 209
dietary, 9, 61, 132, 133, 135, 137, 152, 156, 183, 227
dietary habits, 133
dietary intake, 61, 137
disorder, 3, 10, 15
dispersion, 105, 138
distilled water, 92
distribution, ix, 16, 26, 27, 41, 50, 60, 77, 78, 115, 131, 156, 162, 163, 169, 174, 197, 208, 232, 233, 234
DNA, 10, 20, 31, 32, 33, 45, 135, 157, 171, 199, 203, 214, 236, 266
DNA damage, 20, 31, 45, 135, 157, 171, 199, 203, 236, 266

DNA strand breaks, 32, 157
dominance, 59, 60, 209
Dominican Republic, 14, 178
drinking water, 48, 50, 61, 82, 151, 212
dyspnea, 10

E

earthworms, 160, 173
ecosystem, 15, 78, 150, 160, 180, 183, 197, 205, 207, 212, 213, 216, 217, 238, 251, 252
ecotoxicity, 21, 141, 142, 143, 145, 159, 162, 172
ecotoxicological, 40, 58, 77, 173
ecotoxicology, ix
egg, 33, 47, 67, 260
Egypt, 90, 103, 105, 111, 115, 119, 123, 124, 131, 134, 197, 201
election, 248
electrocardiogram, 31
electron, 82, 85, 129, 188, 219, 264
emission, 4, 199, 209, 233
endocrine, 10, 11, 22, 31, 84, 115, 128, 137, 138, 143, 182, 207, 229, 248, 249, 266, 270
endocrine disruptor, 10, 128, 143, 182
endocrine-disrupting chemicals, 115, 138, 266
endosulfan can be eliminated in agriculture and other sectors, 194
endosulfan diol, 6, 13, 33, 54, 83, 85, 87, 88, 89, 90, 92, 95, 96, 98, 99, 101, 102, 103, 113, 116, 117, 118, 119, 120, 121, 122, 123, 124, 125, 145, 151, 157, 193, 208, 221, 224, 255, 256, 257, 258, 259, 260, 263
endosulfan ether, 83, 85, 87, 89, 90, 95, 96, 98, 99, 101, 102, 103, 104, 114, 116, 117, 118, 119, 120, 121, 122, 123, 124, 125, 145, 157, 159, 179, 193, 254, 255, 256, 257, 258, 259, 260, 261
endosulfan hydroxy ether, 83, 145, 158, 254
endosulfan hydroxycarboxylate, 145
endosulfan hydroxycarboxylic acid, 6, 83
Endosulfan in Pakistan, 179, 182, 184
endosulfan lactone, 6, 83, 85, 87, 89, 90, 95, 96, 98, 99, 101, 102, 103, 104, 106, 114, 116, 117, 118, 119, 120, 121, 122, 123, 124, 125, 145, 151, 158, 224, 254, 255, 256, 257, 259, 260, 263
endosulfan sulfate, 3, 4, 5, 6, 12, 13, 14, 26, 27, 33, 34, 39, 54, 59, 60, 83, 85, 87, 88, 89, 90, 91, 92, 95, 96, 97, 98, 99, 101, 102, 103, 104, 105, 106, 107, 112, 114, 120, 121, 122, 125, 143, 145, 147, 148, 150, 151, 152, 153, 155, 157, 158, 161, 165, 170, 171, 179, 183, 184, 187, 188, 205, 208, 209, 212, 221, 224, 229, 230, 231, 248, 251, 253, 254, 255, 256, 257, 259, 260, 261, 263, 266, 267, 268, 269, 270, 271
Endosulfan α, 85, 87, 88, 89, 90, 91, 92, 93, 94, 95, 96, 97, 98, 99, 100, 101, 102, 103, 104, 105, 106, 107, 116, 117, 118, 119, 120, 121, 122, 123, 124, 125
Endosulfan β, 85, 87, 88, 89, 90, 91, 92, 93, 94, 95, 96, 97, 98, 99, 100, 101, 102, 103, 104, 105, 106, 107, 116, 117, 118, 119, 120, 121, 122, 123, 124, 125
endothelial cells, 214, 235
energy, 32, 206, 208, 221, 249, 252, 254, 260
environment, vii, viii, ix, 1, 4, 6, 11, 12, 14, 15, 18, 19, 20, 21, 23, 24, 25, 26, 27, 29, 30, 32, 34, 38, 40, 41, 42, 43, 45, 47, 48, 50, 53, 54, 55, 64, 71, 72, 73, 76, 78, 79, 81, 83, 115, 126, 128, 138, 141, 142, 143, 145, 148, 152, 160, 161, 162, 163, 165, 168, 169, 170, 171, 173, 174, 176, 178, 182, 194, 196, 197, 200, 201, 202, 205, 206, 207, 208, 212, 213, 216, 227,

229, 233, 234, 235, 236, 247, 249, 265, 266, 273
environmental awareness, 177
environmental conditions, 205, 212, 248
environmental contamination, 77, 176, 180, 198, 212, 229, 238, 265, 267
environmental effects, 186
Environmental Protection Agency, 4, 17, 31, 163, 179, 266
environmental risks, 176
environments, 4, 28, 145, 147, 152, 159, 162, 193
enzymes, 31, 33, 160, 216, 222, 224, 233, 252, 261, 262
epidemiology, 79
equipment, 156, 171
erythrocytes, 32, 158
ester, 25, 27, 83, 106
estrogen, 45, 47, 163
ethanol, 85, 93, 94, 95, 105
ethyl acetate, 85, 97, 105, 106
ethyl alcohol, 91
European Union, 2, 7, 14, 15, 54, 56, 179, 180
evaporation, 27, 28
evidence, 43, 183, 212
examinations, 11
exposure, 2, 4, 9, 20, 25, 30, 31, 32, 33, 34, 43, 44, 47, 49, 63, 82, 83, 111, 113, 126, 127, 129, 130, 132, 134, 135, 136, 137, 138, 139, 152, 153, 156, 158, 166, 167, 171, 172, 176, 183, 187, 197, 198, 199, 210, 222, 227, 232, 235, 250, 251, 266
extraction, 58, 84, 85, 88, 106, 108, 109, 127, 130, 132, 218, 231, 233
extracts, 54, 58, 65, 69, 109, 142, 146

F

farmers, 56, 59, 64, 69, 70, 71, 73, 74, 75, 115, 180, 186, 195, 199, 216
fat, 83, 112, 120, 121, 153, 229
fauna, 22, 29, 141, 151, 154, 249
Federal Register, 17
fertility, 70, 113, 215
fertilizers, 69, 71, 176
Finland, 14, 100, 101, 104, 113, 115, 122, 123, 124, 131, 133, 138, 139, 178
fish, 6, 11, 17, 18, 20, 29, 32, 38, 40, 46, 49, 54, 61, 152, 154, 155, 158, 162, 167, 172, 174, 177, 183, 186, 200, 221, 232, 234
flora, 29, 141, 153, 249
flora and fauna, 29, 249
food, 2, 9, 11, 29, 32, 54, 55, 70, 78, 81, 82, 83, 132, 154, 156, 160, 164, 176, 194, 195, 200, 205, 206, 216, 217
food chain, 11, 29, 32, 81, 83, 205, 216, 217
formation, 151, 158, 160, 166, 211, 214, 215, 220, 221, 248, 251, 253, 260, 261, 262, 263, 268
formula, 4, 5, 58, 83
freshwater, 11, 12, 32, 39, 152, 156, 158, 171, 172, 212, 238
fruits, 2, 54, 55, 142, 177, 187, 188, 200, 201, 209, 210, 214, 248
fungi, 82, 193, 202, 208, 221, 222, 224, 226, 232, 248, 249, 255, 258, 259, 260, 262, 263, 265, 270
furan, 54

G

gas chromatography, 82, 85, 128, 129, 130, 133, 134, 140, 188, 210, 231, 233, 239, 252, 259
gastrointestinal tract, 14
GC/MS, 87, 88, 89, 90, 91, 92, 93, 95, 96, 97, 98, 99, 100, 101, 102, 103, 104, 105, 106, 110
GC/MS/MS, 87, 89, 90, 95, 98, 104, 110
GC–ECD, 103, 110

gel, 96, 100, 101, 102, 103, 105, 109, 263
gel permeation chromatography, 100, 101, 102, 109
genes, 206, 214, 224, 241, 253, 259, 263, 264, 265, 267
genotoxic, 11, 31, 32, 46, 143, 157, 164, 191, 205, 206, 248, 267, 271
germination, 32, 215, 237, 238
Global Environment Facility (GEF), 178
glucose, 31, 158
glutathione, 215
green alga, 12, 23, 151, 222, 240
groundwater, 11, 13, 29, 30, 34, 36, 51, 58, 59, 60, 61, 77, 78, 169, 189, 198, 201, 206, 208, 212, 213, 217, 226, 234
growth, 12, 32, 33, 143, 159, 160, 172, 173, 214, 222, 236, 237, 238, 252, 253, 254, 259, 260, 262
growth rate, 12, 33, 236, 252
guidelines, 58, 110, 126, 150, 162

H

half-life, 4, 6, 13, 27, 28, 54, 154, 207, 248
harmful effects, 32, 71, 186, 194
hazards, 55, 64, 71, 158, 175, 178, 199, 233
health, ix, 2, 4, 9, 11, 15, 16, 44, 54, 56, 63, 64, 73, 74, 76, 79, 80, 82, 132, 138, 141, 142, 145, 152, 157, 171, 180, 186, 194, 195, 198, 207, 210, 226, 232, 238, 266
health and environmental effects, 2, 16
health effects, 9, 54, 79, 152
health impacts, 11, 56, 63, 141, 142, 145
health problems, 4, 157
health risks, 132, 180, 210, 232
hemoglobin, 158
heterotrophic microorganisms, 221
hexane, 85, 87, 88, 89, 90, 91, 92, 93, 94, 95, 96, 97, 98, 99, 100, 101, 102, 103, 104, 105, 106, 108
Hindustan Insecticides, 8, 19, 55

histology, 184, 186
historical data, 155
historical usage and toxicity effects, 176
history, 55, 111, 157
hormone, 135, 137, 141, 157, 250
hormone levels, 135, 137
human, vii, ix, 1, 9, 10, 16, 25, 29, 30, 32, 44, 53, 63, 71, 72, 73, 81, 82, 84, 86, 111, 112, 113, 114, 126, 127, 129, 130, 131, 132, 133, 134, 135, 136, 137, 138, 139, 140, 141, 143, 145, 155, 156, 162, 163, 164, 167, 176, 178, 182, 186, 194, 195, 197, 198, 199, 205, 206, 207, 212, 213, 216, 226, 227, 235, 236, 238, 247, 249, 250, 265, 266, 267
human biomonitoring, 82, 84, 139
human body, 83, 250
human exposure, 82
human health, 1, 25, 53, 71, 72, 73, 81, 126, 145, 156, 162, 176, 182, 194, 195, 197, 199, 206, 207, 212, 216, 226, 247, 250, 265
human leukemia cells, 236
human milk, 127, 131, 132, 133, 227
human samples, ix, 63, 82
human tissues, 82, 186, 227
humidity, 28, 29, 145
hydrogen peroxide, 243
hydrolysis, 4, 13, 34, 145, 208, 224, 240, 261

I

ice pack, 11
Iceland, 14, 178
ideal, 111
identification, 110, 111, 206, 240, 241, 243
illegal, 145, 162, 206, 213
immune response, 235
immune system, 10, 31, 33, 47, 213
immunosuppression, 2, 31, 33

impacts, vii, viii, ix, 10, 11, 20, 29, 32, 33, 34, 53, 54, 61, 63, 141, 156, 160, 173, 179, 195, 201, 213, 215, 216, 247
implemented, 70, 72, 176, 181
imports, 73
in vitro, 208, 232, 248, 266, 270
increasing use of pesticides/insecticides, 194
incubation period, 193, 253, 254, 259, 260
Indonesia, 2, 3, 14, 57, 103, 123, 152, 178
induction, 164, 213, 215
infants, 112, 118, 131, 132, 134, 138, 139, 158
inferences, 158
infertility, 213, 235
infrared spectroscopy, 259
ingestion, 30, 83
insecticide(s), v, ix, 2, 4, 7, 8, 9, 12, 18, 20, 21, 22, 25, 26, 27, 32, 38, 39, 41, 45, 46, 48, 50, 53, 72, 77, 78, 81, 83, 131, 142, 149, 159, 160, 162, 164, 166, 172, 181, 182, 186, 188, 190, 191, 194, 200, 201, 202, 203, 205, 206, 208, 216, 226, 229, 230, 231, 237,238, 241, 242, 244, 245, 247, 248, 249, 265, 267, 268, 269, 270, 271
insects, 2, 11, 26, 54, 69, 82, 143, 177, 182, 183, 205, 207
institutions, 73, 74
invertebrates, 6, 32, 143, 147, 217
ions, 85, 110, 211
iron, 149, 218, 220, 239, 243
irrigation, 150, 176, 189, 198
Islamabad, 175, 184, 186
isolation, 84, 163, 165, 228, 240, 260, 268, 271
isomers, ix, 4, 6, 12, 13, 15, 26, 27, 54, 59, 60, 80, 83, 142, 143, 144, 148, 150, 151, 153, 155, 156, 157, 161, 166, 167, 170, 178, 192, 193, 230, 231, 235, 239, 242, 243, 248, 249, 252, 259, 267

J

Jamaica, 172
Japan, 1, 14, 93, 95, 105, 111, 119, 124, 132, 171, 178
Jordan, 5, 106, 114, 124, 132
juveniles, 32

K

Kazakhstan, 14, 178
kerosene, 6, 179
kidney, 31, 32, 158, 162, 188, 213, 248
kidneys, 234
kill, 32, 69, 83
kinetic studies, 218
kinetics, 149, 219, 220, 230, 239, 243, 252, 253, 259
Korea, 14, 57, 178

L

laboratory studies, 143, 265
lactate dehydrogenase, 214
lagoons, 152
lakes, 23, 150, 152, 177
leaching, 13, 54, 148, 150, 201, 210
lead, 9, 10, 55, 160, 191, 198, 210
life cycle, 16, 233
liquid chromatography, 133, 189, 210
liquid-liquid extraction, 58, 84, 85, 218
liver, 31, 33, 47, 158, 162, 188, 248
local community, 139, 198, 199
local government, 63
low temperatures, 28
luteinizing hormone, 157
lymphocytes, 31, 157, 235

M

magnesium, 32, 218
magnitude, 149, 150, 178
majority, 14, 83, 112, 141, 177, 179
mammals, 3, 13, 40, 41, 143, 162, 183, 207, 213, 250
management, ix, 54, 56, 64, 69, 72, 73, 74, 76, 79, 145, 170, 180, 182, 194, 206, 209, 213, 226, 247, 259, 265
manganese, 149
manipulation, 108
manufacturing, 8, 9, 55, 56
manufacturing companies, 55, 56
manure, 71
marine environment, 161, 169
mass, 75, 82, 85, 128, 129, 130, 132, 133, 134, 137, 140, 210, 229, 231
mass spectrometry, 82, 85, 128, 129, 130, 132, 133, 134, 140, 210, 229, 231
materials, 108
matrix, 85, 109, 111, 115
matter, iv, 73, 149
measurement, 84, 86, 103, 211
measurements, 83, 228
meat, 188, 197, 201
media, 34, 36, 38, 82, 147, 152, 154, 192, 252, 260
Mediterranean, 15, 24, 130, 131, 173, 265
melting, 6, 27
membranes, 156, 242
memory, 2, 10, 31
mental retardation, 2, 10, 31
metabolic intermediates, 252
metabolism, 22, 29, 33, 47, 158, 167, 172, 221, 224, 241, 263, 269
metabolite, 10, 27, 28, 29, 33, 39, 83, 143, 145, 148, 154, 155, 158, 162, 170, 171, 221, 224, 229, 230, 231, 248, 253, 254, 261, 267, 268, 269, 270

metabolites, 20, 26, 45, 48, 60, 82, 83, 84, 85, 109, 110, 111, 112, 113, 114, 115, 116, 117, 118, 119, 120, 121, 122, 123, 124, 125, 126, 127, 129, 132, 134, 140, 145, 153, 157, 160, 161, 168, 171, 239, 240, 249, 252, 254, 255, 256, 257, 258, 259, 266, 267, 269, 270
metabolized, 157, 221
metabolizing, 263, 268
methylene blue, 211, 233
Mexico, 2, 88, 89, 93, 94, 97, 106, 112, 118, 125, 127, 131, 134, 135, 152, 169, 172
microbial community, 205
microorganisms, 12, 82, 160, 168, 205, 208, 216, 221, 222, 223, 224, 225, 226, 228, 230, 239, 240, 241, 268
mineralization, 32, 163, 208, 222, 240, 242, 261, 269
miniaturization, 84
misuse, 75, 177, 181
moisture, 149, 222, 225, 226, 242, 268
moisture content, 225, 226
molecular weight, 5, 83, 144, 218
molecules, 148
mollusks, 11, 155
monooxygenase, 241, 248, 253, 263, 269
Montenegro, 134
mRNA, 263
MTBE, 89, 94, 96
multiplication, 215
mussels, 6, 18, 32, 40

N

National Academy of Sciences, 45, 138
National Institute for Occupational Safety and Health, 148
national parks, 28, 42
National Research Council, 51
negative impacts, 2, 54, 145, 162, 212
nematode, 260

neonates, 118
nervous system, 31, 207, 217
neurological disease, 156
neuronal cells, 44, 156
neurotoxic, 31, 143, 158, 183, 186, 205, 217, 248, 250
neurotoxicity, 43, 45, 54
neurotransmitter, 10
nitrification, 32, 160
nitrogen, 100, 160, 236, 252, 266
non-biological samples, 189
non-polar, 6, 110
North America, 15, 16, 23, 41, 83, 163
nutrients, 70, 160, 173, 238
nutritional status, 216

O

occupational, 2, 10, 20, 43, 44, 82, 113, 126, 148, 152, 156, 171, 177, 196, 198
oceans, 152
octanol-air partition coefficient, 144
octanol-water partition coefficient, 144
optical density, 211
optimization, 128, 206
organic chemicals, 153
organic compounds, 133
organic matter, 127, 148, 160, 168, 190, 197, 202, 216, 219
organism, 3, 38, 59, 255, 256, 257, 258, 271
organochlorine, ix, 2, 12, 15, 16, 17, 18, 19, 23, 25, 26, 27, 38, 39, 40, 41, 47, 48, 49, 50, 51, 53, 59, 60, 76, 77, 78, 79, 82, 113, 127, 128, 129, 130, 131, 132, 133, 134, 135, 136, 137, 138, 139, 140, 150, 152, 155, 163, 164, 166, 167, 168, 169, 174, 178, 186, 187,195, 197, 206, 207, 209, 210, 212, 216, 218, 227, 228, 229, 230, 231, 232, 233, 234, 235, 240, 241, 242, 243, 265, 267, 269, 270, 271
organochlorine compounds, 23, 130, 131, 133, 135, 137
organs, 111, 162, 188
oxidation, 3, 4, 12, 160, 208, 218, 219, 224, 239, 263
oxidative damage, 267
oxidative stress, 135, 136, 213, 214, 226, 233
oxygen, 32, 215, 225, 239, 262
oxygen consumption, 32
ozonation, 218, 239

P

Pacific, 11, 16, 32, 41, 164
Pakistan, viii, ix, 3, 14, 57, 91, 111, 119, 139, 175, 177, 178, 179, 180, 182, 183, 184, 185, 186, 187, 188, 189, 190, 191, 192, 193, 194, 196, 197, 198, 199, 200, 201, 202, 203
paralysis, 2, 217
parasite, 67
parasites, 67
parathyroid, 248
pathology, 111, 113, 234
pathophysiological, 47
pathway, 9, 18, 146, 214, 221, 240, 244, 253, 254, 270
pathways, 18, 153, 230, 262, 271
Pearl River Delta, 234
permeability, 12, 60, 266
persistence, 2, 4, 15, 60, 64, 83, 142, 149, 191, 202, 208, 226, 228
persistent, ix, 1, 14, 15, 16, 18, 19, 20, 21, 23, 24, 25, 26, 34, 38, 39, 40, 42, 43, 46, 47, 49, 50, 51, 53, 54, 55, 56, 60, 64, 76, 77, 78, 81, 83, 114, 115, 127, 128, 129, 131, 133, 137, 138, 143, 174, 175, 176, 177, 178, 179, 194, 196, 199, 205, 206, 207, 209, 221, 226, 227, 228, 234, 235, 249, 259, 261, 273, 274

persistent organic pollutant (POPs), 1, 8, 14, 16, 19, 23, 26, 28, 34, 38, 39, 42, 43, 46, 47, 49, 55, 73, 76, 80, 83, 127, 128, 131, 137, 138, 175, 176, 177, 178, 180, 194, 199, 206, 207, 227, 228, 259, 261, 273, 274

pests, 2, 7, 55, 64, 69, 71, 82, 142, 145, 148, 170, 175, 176, 182, 186, 195, 206, 209, 216, 248

petroleum, 86, 103, 106

pH, 5, 6, 13, 89, 92, 97, 151, 192, 193, 218, 219, 220, 222, 225, 226, 252, 253, 259, 260, 261

Philippines, 3, 7, 14, 18, 57, 177, 178, 212, 234

phosphate, 31, 92, 97, 217

phosphorus, 217

photocatalysis, 218

photochemical degradation, 239

photooxidation, 208, 267

physicochemical characteristics, 2

phytoplankton, 12, 22, 61, 229

placenta, 82, 85, 86, 100, 101, 110, 113, 114, 122, 126, 127, 131, 133, 138, 139

plant growth, 153, 173, 215, 237, 269

plants, 7, 12, 13, 20, 28, 32, 69, 147, 149, 153, 170, 208, 209, 215, 231, 236, 248

plasma, 14, 82, 84, 85, 87, 91, 92, 97, 111, 116, 117, 118, 119, 130, 135, 219, 220, 242, 243

poisoning, 2, 9, 10, 11, 12, 15, 17, 30, 31, 39, 43, 44, 55, 63, 128, 141, 157, 177, 178, 179, 227, 266

pollutant, 1, 8, 14, 16, 19, 23, 26, 28, 34, 38, 39, 42, 43, 46, 47, 49, 55, 73, 76, 80, 83, 114, 127, 128, 131, 137, 138, 152, 175, 176, 177, 178, 180, 194, 199, 206, 207, 213, 214, 216, 227, 228, 248, 259, 261, 273, 274

polybrominated diphenyl ethers, 40, 48

polychlorinated biphenyl, 39, 40, 41, 48, 77, 132, 133, 138, 166, 227

polychlorinated biphenyls (PCBs), 132

polycyclic aromatic hydrocarbon, 170

Portugal, 81, 90, 100, 111, 112, 119, 121, 135, 209

possible measures, 176

postnatal exposure, 138

potato, 153, 170, 184, 187

preparation, 82, 85, 101, 108, 130, 132, 133, 180, 229

preservative, 2, 54, 177

preterm delivery, 117, 136

prior knowledge, 180

production, 2, 7, 8, 9, 26, 54, 55, 56, 69, 70, 71, 83, 114, 141, 142, 151, 160, 175, 176, 179, 180, 181, 195, 206, 210, 226, 235, 254, 260

production costs, 71

protection, 54, 65, 66, 72, 73, 74, 152, 153, 156, 162, 200, 202

Pseudomonas aeruginosa, 145, 192, 203, 232, 249, 251, 253, 256, 268, 269

public health, 54, 71, 73, 74, 81, 83, 168, 188

Q

quantification, 84, 110, 136, 210, 233

R

rainfall, 11, 16, 189

reaction mechanism, 239

reactions, 18, 145

receptor, 33, 45, 156

recovery, 85, 126, 242

Registry, 42, 158, 167, 172, 266

regulations, 72, 177, 180, 206

relief, 79

remediation, ix, 79, 192, 194, 208, 237, 247, 251, 253, 261, 265, 268

reproduction, 33, 160, 173

reproductive age, 113
reproductive organs, 207, 213
research institutions, 194
researchers, 34, 157, 261
residue, 12, 29, 54, 63, 131, 136, 141, 145, 150, 151, 154, 161, 168, 187, 188, 189, 197, 201
residue levels, 131, 136, 141, 145, 150, 151, 161, 188
residues, 6, 18, 22, 34, 36, 38, 39, 40, 42, 48, 49, 50, 51, 53, 59, 60, 76, 77, 78, 79, 83, 110, 111, 128, 129, 130, 131, 132, 133, 134, 136, 138, 139, 141, 147, 149, 152, 153, 154, 155, 157, 161, 162, 169, 170, 172, 174, 186, 187, 188, 190, 195, 198, 199, 200, 201, 224, 227, 228, 229, 232, 233, 234, 248, 265
resistance, 46, 81, 126, 145, 182, 185, 186, 190, 193, 202, 203, 212, 237
resolution, 110
resources, 69, 79, 171, 196
respiration, 10, 160, 173, 215
risk, ix, 16, 17, 38, 39, 44, 50, 54, 55, 58, 61, 62, 77, 78, 81, 82, 113, 129, 133, 138, 139, 152, 155, 171, 176, 197, 198, 199, 206, 213, 216, 234, 266
risk assessment, 17, 39, 54, 55, 62, 77, 78, 133, 155, 171, 197, 198, 234
risk factors, 138, 266
risk perception, 199
risk profile, 16, 38
risks, 44, 155, 156, 169, 173, 176, 195, 197, 199, 206, 212, 234
roots, 70, 153, 215
Rotterdam Convention, 14
routes, 30, 75, 150, 152, 221
runoff, 13, 54, 150, 152, 168, 174, 206, 212, 217

S

safety, 44, 77, 180, 198, 199, 206, 226, 237, 243, 265
sample preparation, 82, 84, 85, 108, 130, 132, 133
science, ix, 40, 41, 76, 78, 197, 198, 202, 227, 233, 239, 241, 267
scientific investigations, 145
sediment, 16, 27, 34, 36, 40, 63, 77, 127, 131, 150, 152, 155, 162, 168, 173, 174, 207, 229
sediments, 11, 20, 26, 27, 29, 34, 36, 38, 40, 41, 50, 54, 77, 82, 147, 148, 150, 151, 152, 162, 169, 190, 221, 239, 248
seed, 54, 67, 69, 215, 238
seedlings, 70, 215
seizure, 10, 158
semi-volatile, ix, 4, 12, 25, 26, 27, 142, 207, 210
sensitivity, 126, 155
Serbia & Montenegro, 14, 178
serum, 82, 84, 85, 87, 88, 89, 90, 95, 96, 111, 116, 118, 119, 128, 129, 130, 131, 132, 133, 134, 135, 137, 157, 158
services, 44, 75, 238
sex, 137, 157, 250
shoot, 65, 66, 68, 153, 214
showing, 115, 250, 265
signaling pathway, 214, 235, 236
silica, 58, 88, 90, 93, 94, 95, 96, 99, 100, 101, 102, 103, 105, 108, 110, 193
Singapore, 14, 44, 178
sludge, 206, 247, 261
sodium, 92, 93, 95, 104, 106, 149
solid phase, 58, 88, 129, 189, 218, 233
solid phase microextraction, 108, 129, 189
solid-phase extraction, 84, 108, 109, 127, 130, 132
solubility, 6, 18, 27, 33, 81, 144, 217, 221, 224

solution, 86, 89, 90, 92, 192, 193, 211, 239, 242
sorption, 29, 168, 193, 201, 207, 219, 221
sorption coefficient, 33, 144
sorption method, 193
South Africa, 11, 21, 43, 93, 111, 119, 130, 233, 234
South America, 7, 15, 24, 33, 83
South Asia, ix, 180, 248
South Korea, 3, 9, 228
South Pacific, 15, 24, 196
Southeast Asia, 9
Soviet Union, 8
Spain, 87, 88, 89, 90, 95, 96, 98, 99, 101, 102, 103, 104, 111, 112, 113, 114, 115, 116, 120, 122, 123, 125, 130, 134, 135, 136, 137, 138, 139
species, ix, 11, 21, 29, 38, 54, 59, 61, 108, 145, 149, 154, 155, 166, 192, 208, 215, 222, 224, 227, 230, 237, 240, 241, 242, 248, 251, 252, 260, 261, 262, 263, 270, 271
specifications, 17
spermatogenesis, 214
Sri Lanka, ix, 3, 14, 17, 39, 57, 141, 178, 227
stability, 5, 13, 64, 209, 214, 239
state, 18, 38, 49, 56, 63, 69, 70, 72, 76, 139, 157, 178, 227
states, 58, 59, 64, 70, 73, 78
Stockholm Convention, 1, 8, 14, 26, 42, 55, 56, 72, 73, 83, 164, 165, 175, 177, 180, 182, 194, 207, 227, 228
Stockholm Convention on Persistent Organic Pollutants, 56
storage, 56, 74, 159
stress, 214, 215, 224, 236, 237
structure, 15, 27, 28, 32, 224
sub-Saharan Africa, 238
substrate, 109, 111, 208, 221, 224, 229, 251
sub-tropical, 161, 162

sulfate, 3, 4, 5, 6, 12, 13, 14, 26, 27, 33, 34, 39, 54, 59, 60, 82, 83, 85, 87, 88, 89, 90, 91, 92, 95, 96, 97, 98, 99, 101, 102, 103, 104, 105, 106, 107, 111, 114, 120, 121, 122, 125, 143, 145, 147, 148, 150, 151, 152, 153, 155, 157, 158, 161, 165, 170, 171, 179, 183, 184, 187, 188, 205, 208, 209, 212, 219, 221, 224, 229, 230, 231, 236, 248, 251, 253, 254, 255, 256, 257, 259, 260, 261, 263, 266, 267, 268, 269, 270, 271
survival, 11, 21, 183, 194
survival rate, 11
survivors, 10
susceptibility, 183, 190
sustainable development, 79
Sustainable Development, 247
synergistic effect, 221
synthesis, 157, 208, 209, 214, 237, 250, 259

T

Tanzania, 107, 115, 125, 129, 132
target, 12, 20, 22, 32, 45, 46, 85, 109, 110, 148, 161, 195, 207
technical endosulfan, 4, 6, 10, 12, 148, 165
techniques, ix, 58, 71, 82, 84, 86, 87, 98, 100, 103, 110, 132, 189, 193, 201, 208, 210, 218, 223, 239, 241, 247, 249, 251, 264, 268
technology, 40, 70, 76, 176, 218, 226, 233, 239, 267
temperate, 3, 141, 143, 145, 147, 149, 161, 162
temperature, 28, 29, 43, 87, 145, 191, 192, 193, 218, 222, 242, 252, 254, 260
testosterone, 10, 157
Thailand, 3, 14, 17, 57, 150, 168, 178
the implementation stage of such legislations/regulations is poor in pakistan, 180

thin films, 13
tissue, 31, 32, 86, 112, 113, 156, 162, 174
tobacco, 2, 9, 54, 177, 189
toluene, 100, 101, 104, 105
total product, 8, 26
toxic effect, 158, 177, 183, 186, 249, 262
toxic substances, 44, 175, 194, 238
toxicity, viii, 10, 11, 14, 17, 20, 21, 22, 25, 26, 30, 31, 45, 46, 47, 48, 54, 64, 82, 141, 142, 143, 145, 156, 158, 159, 168, 172, 173, 176, 177, 179, 182, 183, 190, 194, 206, 207, 208, 213, 216, 217, 221, 224, 229, 234, 235, 237, 244, 248, 250, 252, 263, 265, 267
toxicity assessment, 190, 194
toxicology, 39, 44, 77, 137, 139, 177, 229, 238, 265, 266, 267
trade, 3, 57, 59, 72, 142
training, 73, 74, 75, 181
transformation, 6, 12, 141, 143, 145, 147, 160, 162, 250, 253, 263
transformation product, 6, 141, 143, 145, 147, 162
transport, 1, 4, 12, 15, 16, 19, 25, 26, 28, 39, 40, 41, 61, 72, 142, 152, 163, 166, 232
transport processes, 12
treaties, 73
tropical, viii, ix, 16, 41, 141, 142, 145, 147, 148, 152, 156, 159, 160, 161, 162, 164, 165, 166, 167, 169, 173
tropical environment, 41, 142, 145, 148, 162
tumor necrosis factor, 213

U

umbilical cord, 82, 87, 97, 112, 116, 118, 126, 130, 137
umbilical cord blood, 82, 97, 116, 126, 130, 137
United Nations, 2, 17, 19, 20, 21, 23, 24, 26, 42, 79, 142, 178, 195, 196

urban, 40, 48, 166, 187, 195, 207, 209, 228, 232
urban population, 187
urine, 69, 82, 84, 85, 115, 125, 127, 139, 140
US Department of Health and Human Services, 16
USA, 12, 26, 42, 95, 111, 119, 129, 172, 238, 266
UV irradiation, 208, 219

V

validation, 133
vapor, 4, 6, 13, 27, 142, 218
vapor pressure, 4, 5, 6, 27, 33, 142, 144, 218
variations, 39, 46, 47, 49, 76, 147, 165, 210, 231, 233
varieties, 237, 261, 262
vegetables, 2, 13, 54, 142, 177, 187, 191, 198, 200, 206, 209, 210, 214, 216, 226, 248
vegetation, 4, 26, 27, 29
volatilization, 4, 6, 28, 29, 54, 142, 145, 147, 149, 150, 154, 167
vomiting, 10

W

waste, 70, 194
waste water, 194
water quality, 155
water solubility, 27, 33, 144
weight gain, 159
West Africa, 2, 44, 149, 151, 164, 170
wildlife, 12, 207
workers, 31, 108, 115, 116, 119, 185, 186, 200, 224, 259, 263
World Health Organization (WHO), 54, 179
worldwide, 2, 7, 25, 55, 115, 122, 176, 177, 178, 207, 213

X

xenon, 218

Y

Yaounde, 238
yeast, 217
yield, 70, 209

Z

zeolites, 218, 238

zinc, 239
zooplankton, 12, 22, 61, 229

α

α- endosulfan, 6, 28, 145, 258, 259

β

β-isomer, 6, 12, 142, 145, 151, 153, 156, 157, 217, 221

Related Nova Publications

Managing Stormwater: Practices and Challenges for Reuse and Recycling

Editor: A.H.M. Faisal Anwar, PhD

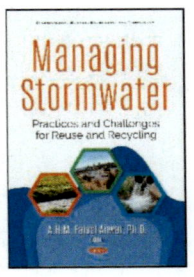

Series: Environmental Science, Engineering and Technology

Book Description: This book has eleven chapters that describe the practices and challenges of different BMPs for stormwater management. These include combined sewer networks, different rainwater harvesting techniques, constructed wetlands, MUSIC modelling of bioretention systems, catch basin inserts, permeable pavements, the use of adsorbents for cleaning stormwater, low impact developments, and membrane-based technologies for stormwater treatment.

Hardcover ISBN: 978-1-53615-250-0
Retail Price: $230

Liquefaction: Analysis and Assessment

Editor: Corey Hanson

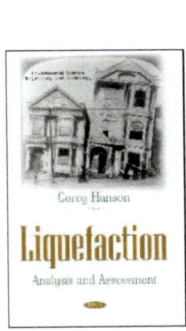

Series: Environmental Science, Engineering and Technology

Book Description: *Liquefaction: Analysis and Assessment* introduces hydrogen liquefaction techniques and provides an outlook concerning common methodology. Basic and conventional cycles used for hydrogen liquefaction are reviewed, and industrial in-service hydrogen liquefaction plants are discussed.

Hardcover ISBN: 978-1-53614-773-5
Retail Price: $160

To see a complete list of Nova publications, please visit our website at www.novapublishers.com

Related Nova Publications

PHOTODEGRADATION: MECHANISMS AND APPLICATIONS

EDITOR: Frank Soto

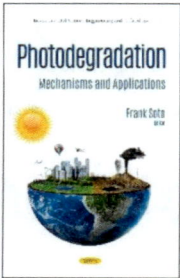

SERIES: Environmental Science, Engineering and Technology

BOOK DESCRIPTION: Organic pollutants are harmful to the ecological environment and human health. In *Photodegradation: Mechanisms and Applications,* the authors propose photodegradation as an attractive strategy to remove organic pollutants. Photodegradation only requires suitable photocatalysts and light, and the sun is an inexhaustible light source.

HARDCOVER ISBN: 978-1-53614-568-7
RETAIL PRICE: $160

STRATEGIC ADVANCES IN ENVIRONMENTAL IMPACT ASSESSMENT: CHALLENGES OF UNCONVENTIONAL SHALE GAS EXTRACTION

AUTHOR: Afsoon Moatari Kazerouni

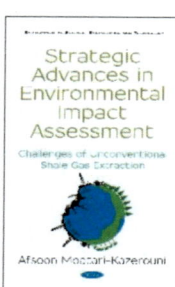

SERIES: Environmental Science, Engineering and Technology

BOOK DESCRIPTION: Shale gas is natural gas that is tightly locked within low permeability sedimentary rock. Recent technological advances are making shale gas reserves increasingly accessible and their recovery more economically feasible.

HARDCOVER ISBN: 978-1-53614-433-8
RETAIL PRICE: $195

To see a complete list of Nova publications, please visit our website at www.novapublishers.com